Praise

Pure Poultry

Part memoir and part roadmap to success, *Pure Poultry* tells a compelling story full of grit, passion and discovery as the author and her husband embrace a sustainable, off-grid life in the woods — with chickens, turkeys and ducks. Whether you are a seasoned fowl keeper or a passionate dreamer, you will find yourself nodding, laughing, commiserating, learning and simply enjoying the author's knack for weaving her experiences, observations and lessons into a page-turner of a narrative. Once I started reading, it was next to impossible to put *Pure Poultry* down, even when I heard my own geese honking for their supper. And when I finally closed the book, for the last time, I knew exactly why I keep the fowl that I do, I learned some tricks I'd not discovered in the past 35 years and I felt happy!

— Oscar H. "Hank" Will; sustainable farmer;
editor in chief, *GRIT Magazine*; author, *Plowing with Pigs*

In *Pure Poultry*, [Miller's pragmatism and mirth] come through in a beautiful blend, making this a great book for anyone who is dreams of waltzing down the country-living path (whether or not you intend to raise heritage poultry). And in sharing her own life experiences, she also provides an excellent resource for the aspiring poultry person to learn the how to's of keeping chickens, ducks and turkeys in your own backyard or on your own farm.

— Carol Ekarius, author, *Storey's Illustrated Guide to Poultry Breeds*

Pure Poultry is not merely about chickens, ducks, and turkeys, but about one couple's off-grid sustainable life style that includes keeping free-range poultry for fun and profit. Author Victoria Redhed Miller's engaging style draws you in right from the start and holds you captive all the way to the end. Reading *Pure Poultry* is like visiting with an old friend.

— Gail Damerow, author of *The Chicken Encyclopedia*

Pure Poultry is a fun, easy read and full of personal stories. Sharing their experience raising turkeys, ducks and chickens is a lesson to all. All beginners can learn from their wins and mistakes at getting started with poultry.

— Frank R. Reese founder of Good Shepherd Poultry Ranch

pure
POULTRY

Living Well *with*
HERITAGE CHICKENS, TURKEYS *and* DUCKS

Victoria Redhed Miller

new society
PUBLISHERS

Cover design by Diane McIntosh.
Photo by Victoria Redhed Miller.
All interior photos by Victoria Redhed Miller, unless otherwise noted.

Printed in Canada. First printing August 2013.

Paperback ISBN: 978-0-86571-753-4
eISBN: 978-1-55092-546-3

Inquiries regarding requests to reprint all or part of *Pure Poultry* should be addressed to New Society Publishers at the address below.

To order directly from the publishers, please call toll-free (North America) 1-800-567-6772, or order online at www.newsociety.com

Any other inquiries can be directed by mail to:

New Society Publishers
P.O. Box 189, Gabriola Island, BC V0R 1X0, Canada
(250) 247-9737

New Society Publishers' mission is to publish books that contribute in fundamental ways to building an ecologically sustainable and just society, and to do so with the least possible impact on the environment, in a manner that models this vision. We are committed to doing this not just through education, but through action. The interior pages of our bound books are printed on Forest Stewardship Council®-registered acid-free paper that is **100% post-consumer recycled** (100% old growth forest-free), processed chlorine free, and printed with vegetable-based, low-VOC inks, with covers produced using FSC®-registered stock. New Society also works to reduce its carbon footprint, and purchases carbon offsets based on an annual audit to ensure a carbon neutral footprint. For further information, or to browse our full list of books and purchase securely, visit our website at: www.newsociety.com

Library and Archives Canada Cataloguing in Publication

Miller, Victoria Redhed, author
        Pure poultry : living well with heritage chickens, turkeys and ducks / Victoria Redhed Miller.

Includes bibliographical references and index.
Issued in print and electronic formats.
ISBN 978-0-86571-753-4 (pbk.).--ISBN 978-1-55092-546-3 (Ebook)

        1. Chickens. 2. Turkeys. 3. Ducks. I. Title.

SF487.M54 2013   636.5   C2013-905086-8
                                                     C2013-905087-6

Contents

Foreword

Jeannette Beranger
Research & Technical Programs Manager
American Livestock Breeds Conservancy

THE FIRST TIME I MET VICTORIA MILLER AND HER HUSBAND DAVID, it was on a gray and cold day in January of 2009. I received a warm welcome from both, along with their flock of Midget White turkeys. I had been invited to tour their farm and along with the Millers was escorted by the turkeys who followed us everywhere, ensuring that they were not excluded from anything interesting going on. The tour quickly melted the chill away and brought me into the Millers's marvelous world of homesteading, which this book is sure to do for its readers.

The trip to the Millers's Canyon Creek Farms occurred by way of traveling to Washington state and taking an extra day or two from my work to do some "farm hopping" and visit with members of the American Livestock Breeds Conservancy (ALBC) working with endangered breeds on their farms. I was especially interested in what was happening on the Olympic Peninsula since farming can be a challenge with the wet climate, large predators, and general remoteness of the region.

As I talked and toured with Victoria, I learned that despite all of these challenging factors, the Millers have managed to find a way to make things work around the farm. Using traditional breeds of poultry was paramount to their goals which in turn support genetic diversity for the species through breeding and sharing with others.

The successes are a testament to stick-to-itiveness and determination to be kind to the land and to make a difference with the farm. Victoria and David's methods are simple, many developed through trial by fire as you will read, but they worked.

At the American Livestock Breeds Conservancy, we are aware of the need for more sources of information on living with heritage breeds. *Pure Poultry* comes directly from the Millers's experiences raising chickens, turkeys and ducks, starting out as beginners with six hens and a rooster. The day-to-day interaction between the birds and their caretakers reveals not only the advantages of heritage breeds, such as foraging ability, but also their different, entertaining, and often quirky personalities.

Pure Poultry is not just another how-to book about raising poultry. As you follow the story of the Millers's first few years with their birds, it will help you learn whether raising heritage poultry might work in your particular situation. Included is a detailed worksheet designed to walk you through the process of making the right choices, and to ensure you and your family are prepared to begin raising poultry.

In this book, you will experience the humble wisdom of the Millers that makes homesteading with heritage breeds successful and how "simple" can be an exceptionally rewarding lifestyle. Sit back, enjoy, and learn!

Introduction: (Heritage) Chickens and Turkeys and Ducks: Oh, My!

A BOUT A HUNDRED AND FIFTY YEARS AGO, when Abraham Lincoln was president, 48 percent of the American population was farming, and nearly everyone lived somewhere close to fresh food sources. Currently, less than 2 percent of our population raises our food.

Think about what that means. We are that dependent on someone else to feed us.

The development of the interstate highway system, the advent of air freight and a major shift in the population toward big cities — all these factors have contributed to significant changes in what we eat. Most of us live so far away from these farms and processing plants that more and more fossil fuels are being used just to transport the food to us. How good is the quality of that food by the time it reaches our local grocer and, later still, our tables? And how much money does the farmer pocket, once all the middlemen take their cut?

Sustainability: What It Means, Why It Matters

The dictionary definition of *sustainable* is, in part: "a method of harvesting or using a resource so that the resource is not depleted or permanently damaged."

Wow. Suppose we could simplify the way we feed ourselves, and do it sustainably? Suppose that, by choosing to grow some of our food, we also contribute beauty and a healthier environment to our world? Suppose that, in the process, we find a richer, more self-sufficient life, as individuals, families, communities — even at a national or global level?

What do you actually know about where your food comes from? How was the chicken you ate last night raised? You might not want to know; frankly, you might not care. Okay, fair enough. But we can't ultimately have it both ways.

Suppose we start to take back the responsibility for feeding ourselves, and at the same time, nurture our natural environment and help support our local community. I believe that, no matter where you live or how much space or time you have, you can begin to make choices that will, step by step, ultimately result in the satisfaction of being more self-sufficient. It's not just about having the freshest eggs ever or your own homegrown holiday turkey; it's about a mindset that is willing to acknowledge — and pursue — a better way of doing things. In fact, a better way of life.

Why Raise Poultry at All?

Historically, in times of war or a tough economy, Americans have relied on the small family farms to provide their food. Worldwide, there is much less arable land per capita now than there was 100 or even 50 years ago. So where is our food coming from? Increasingly, people are choosing to take up the challenge of producing some of their own food. Besides the obvious benefits of eating the freshest, most nutritious foods at a lower cost, they're discovering the joy and satisfaction of providing for themselves and their families.

This was certainly part of our motivation when my husband David and I began planning how we would live once we moved to the farm. Having gained plenty of experience over the years with gardening and food preservation techniques, I was already familiar with that sense of pride and accomplishment. In addition to growing fruits and vegetables, though, we expected that at some point we would take the plunge and try raising animals. Having heard lots of stories from my mother, who raised pigs while growing up in Illinois, I really wanted to raise them too. At the time, though, David wasn't too enthusiastic. Although initially disappointed, I decided we shouldn't go ahead with pigs unless we were both really excited about it. After a bit more debate, and some initial research, we agreed to start with chickens. Just a few laying hens, we thought, to provide enough eggs for us and maybe some extras to share with friends and family.

Knowing what I know now, I would raise poultry for the entertainment value alone! I wouldn't ever have guessed that there are such different personalities between chickens, turkeys and ducks. Just trying to decipher the different vocalizations of the chickens is wonderfully diverting. The turkeys are incredibly friendly and curious, and just want to follow us around whenever we're outside. Actually, they also walk around the house when we're inside, trying to see where we are and what we're up to. The ducks have an adorable way of looking up at us sideways — I call it the Princess Diana look — and have definite games they play with us, mostly when it's time to close them up for the night. Ducks also look like they're always smiling.

Heritage vs. Hybrid

What does "heritage" mean, as far as poultry goes? It's really just another term for a purebred bird, as distinct from a hybrid. A purebred bird is one that results from the mating of two genetically similar parents. Hybrid varieties are produced from genetically dissimilar parents. Hybrid poultry breeds have, in general, been bred for some specific trait or traits, such as high egg production or rapid maturity. Most of these hybrid poultry breeds were developed relatively recently; there were no hybrid varieties in the poultry industry prior to 1930.

The commercial poultry industry overwhelmingly favors hybrids, due to their comparatively fast growth and high production capabilities. Examples are the familiar Cornish Cross meat chicken and the Broad Breasted turkey (typically the Nicholas strain, Giant White), both of which are commonly found in large grocery stores. However, there are numerous problems that can arise over multiple generations of cross-breeding. The main issue is that, in the process of selecting for just one trait, such as high egg production, the overall health of the bird is ignored. Combined with the typical extreme overcrowding in commercial poultry operations, this has led to the necessity of adding sub-therapeutic doses of medications to the birds' feed in order to offset their lower disease resistance and their unhealthy living conditions. This means that the poultry industry has found a way to ultimately separate their own bottom line from the health and well-being of the birds.

On the backyard or small-farm level, though, heritage-breed poultry have much to offer. Where the goal is sustainability, they win every time over hybrids. Lately a growing number of poultry lovers (and other thoughtful consumers) in America are turning back to the future: discovering the pleasures of raising hardy, disease-resistant birds that mate naturally, incubate, hatch and raise their offspring and forage much of their own food. And that's just the beginning of how beautiful, sustainable heritage poultry can contribute to your more self-reliant lifestyle.

Benefits of Heritage Breeds

One of the most obvious benefits of raising purebred poultry is that they produce offspring that are true to type — a clear advantage over hybrids, which must be replaced by buying more birds. Among the many other reasons that make heritage breeds an attractive choice:

- There are many breeds to choose from, and they are adapted to many different climates.
- They mate naturally and brood their own offspring.
- They happily provide part of their own food by foraging.
- They help control weeds and harmful insects, and even small rodents.
- Their disease resistance and longevity are superior.
- They are easy to care for.

Here at Canyon Creek Farms, the only poultry breeds we raise are heritage breeds. We have no personal experience raising hybrids such as the Cornish Cross meat chicken. You might be thinking that we would change our minds about raising dual-purpose heritage chickens for meat production if we gave the Cornish Cross a chance. I personally haven't met anyone around here who, given the choice, would opt for the beautiful dual-purpose New Hampshire, our choice for farmstead meat and egg production (see Chapter 27). Yes, we have to wait longer for our New Hampshires to grow to a good slaughter size. Sure, the hens don't lay as many eggs per year as production laying birds. However, we are able to breed our birds, saving the expense of buying new chicks every year. Our New Hampshire cockerels, which typically dress out at five pounds or more by about 18 weeks, are a whole lot tastier than the comparatively bland supermarket chicken. And our hens will still be laying their delicious, nutritious eggs long after those commercial layers reach the end of their second year — and the end of their useful laying life.

Depending on your goals, it may turn out that hybrids such as the Cornish Cross make more sense for you: for example, if you aim for a large profit-oriented meat-bird operation that will be your sole source of income. My motivation in writing *Pure Poultry* is not to convince you that heritage-breed poultry is the answer in every imaginable situation. I simply hope to show, through relating our actual

New Hampshire pullets on our old split-rail fence.

experiences in our first few years on the farm, how heritage-breed chickens, turkeys and ducks can contribute to a healthy and satisfying way of life.

We choose heritage breeds for many reasons, but there is no doubt that, compared with hybrids, they are simply a better fit for our way of life.

Chapter 1

When Good People Get Poultry

"**W**AKE UP," my husband David said softly. It was early on a Saturday morning in late May. "It's Chicken Day."

We had been working hard to get ready for the arrival of our first chickens, and the day had come, too quickly and not quickly enough.

I don't remember much of what I did all day, although I'm sure I was fidgeting with the innocent anticipation of a little kid on Christmas Eve. The week before, we had met some people a few miles down the hill from us. They were in the process of moving, downsizing actually. Since there was no room for chickens at their new place, the birds had to go.

The chickens turned out to be Buff Orpingtons. Great! I thought. The Orpington was on my short list of preferred breeds, those that met all my criteria (more about this later).

I realized — too late — that I had failed to pay adequate attention when the birds' owner mentioned an incident of the rooster rushing at him, quite aggressively, for no apparent reason. This bird was large, with a permanently peevish expression. I assumed (blush) that if we were just "nice" to the chickens, the rooster would mellow out.

Not knowing any better, in spite of all my research, we also took at face value the man's claim as to the birds' age. "A year and four months," he confidently stated. Much later, we concluded that if those numbers had been reversed, it would have been closer to the

7

mark. The following spring, when our second-generation roosters grew spurs that were maybe an inch long by their first birthdays, we wondered. Papa Cock's spurs must have been a good four inches long, not that we ever got close enough to measure. At least, not while he was alive.

Anyway, it was Chicken Day. We expected delivery of the birds and their coop late that afternoon. Although it was a bit pricey, we had decided to buy the coop from the owners because it had good storage space for feed and equipment. It also had nest boxes and roost space for more than four times the number of birds we were starting with. Plus, they would deliver it for a small fee.

I was nervous and excited and impatient by the time they drove up. The chickens were in waxed cardboard Chiquita banana boxes, which David and I transferred to the back of our car while the guys were maneuvering the trailer to put the coop in place. Finally, it was done, and the truck drove off as dusk began to deepen over the Olympic foothills.

Although it was a little early, we decided that it would be best to just put the birds into the roost for the night and let them settle down. So we took them, one by one, from their banana boxes and gently placed them into the roost through the side door, talking quietly to them. I was entranced by their soft clucking and cooing as they moved around the roost area. They were just beautiful. This was good.

After they were all tucked in and the coop doors secured, we walked back to the house. Actually I think I floated; it felt like Christmas, at the moment when you look at the pile of paper and ribbons and bows and realize the last gift has been opened. The anticipation of earlier that day had given way to a mixture of feelings: excitement at the start of something new; satisfaction (and relief) that we'd gotten through this part of the process; and even yet a little nervousness, on my part anyway. Had I learned enough? Would we be good caretakers of these beautiful creatures? Would they like it here? Would they be attacked and killed by a swarm of mutant killer bees their first day on our farm?

As we settled in for the night, David said, "Don't be disappointed if the hens don't lay eggs right away. They'll probably be upset by the move; they need a little transition time."

Okay, I thought sleepily. As I dozed off, it occurred to me to wonder how he would know that. Must ask him in the morning.

Chapter 2

Another Beautiful Day in Paradise

CANYON CREEK FARMS wasn't always a farm. David's Portuguese grandparents bought the land in 1936 after the previous owner defaulted on his property taxes. At the time, David's mother, Lorelei, the youngest of the three Moniz sisters, was about nine years old. The five of them settled into the tiny old wooden house that was already there, at the top of a ridge overlooking a grassy meadow to the east and natural peat bog beyond.

The property had been clear-cut back in the 1920s. When the Moniz family arrived, there were not many trees here. Now, more than eighty years later, there are so many trees that it's difficult to imagine the farm without them. Something like thirty-three of our forty acres are wooded, with a mix of several kinds of fir, red alder, cedar, cottonwood, maple and even a few madrone trees. I didn't realize how much of our property — about a third — is down in a canyon until I saw an aerial photo. On the west side of the road, the land drops off, steeply in places, down three hundred feet to Canyon Creek. Since the runoff from both of our ponds goes downhill to the creek, we realized that there is excellent potential for hydroelectric power.

Speaking of power, our farm has none. Our property is two miles off the electrical grid; that is, we don't have full-time electricity. Our house is about two miles up the hill from our nearest neighbor and the end of the utility grid. Frankly, even if we had the money (approximately $100,000; this is not a misprint) to connect to the grid, we would still choose to keep things as they are. As I write this, we are almost finished installing the direct-current (DC) part of our new solar electric system, everything, that is to say, except for the actual house wiring. This system is designed to provide all the power we need for the house, a fully equipped power tool shop and eventually a guest house. For now, though, the house runs on propane and is quite comfortable. We have hot water, a gas refrigerator/freezer, a gas stove and oven and gas lamps on the walls. Two wood stoves provide our heat about eight months of the year. And I love the fact that we have three guest rooms.

In fact, this house, which David's grandfather built in the late 1940s, is larger than the one we lived in in Seattle before we moved here in 2006. It's a wonderful house, built to be fire-resistant (we're surrounded by many, many acres of woods in a fairly dry climate) as well as to withstand mountain weather, including the windstorms and snowfall that pummel us each winter. Naturally we have an abundant supply of wood for fuel; the fir and alder especially make beautiful firewood. Considering the size of the house (about 2,300 square feet), it's amazing how well the two wood stoves heat it. Of course, we may not always be as excited as we are now to be cutting, hauling, splitting and stacking so many cords of firewood every year. It's a lot of work. But we love this place.

Our first year here, 2006, was spent mainly in transition. How I came to hate that word, along with the phrase, "It's only temporary." The sale of our house in Seattle wasn't due to close until early November that year, so we spread out the moving process somewhat. Also, the house we were moving into needed lots of work to make it livable, once the longtime tenant moved out. Junk had to be hauled out and disposed of. New appliances and wood stoves had to be researched, bought, shipped and installed. One room at a time,

I emptied, cleaned and repainted the house. I bought and installed shades and insulating curtains. Chose new handles for the kitchen cupboards (they previously had none). I even located an authentic Hoosier cabinet, something I'd dreamed of but nearly gave up on finding in this part of the country. Over the next few months, as we went back and forth from Seattle, our new home became truly ours.

One of the big projects that first summer was replacing the old propane tank, a 400-gallon above-ground eyesore. Mostly for safety reasons, David wanted an underground tank, so we bought a 1,000-gallon model. We had decided on the location, on a west-facing slope about 50 feet away from the southwest corner of the house. Call us nutty, but we decided from the get-go to dig the hole ourselves, by hand. Uh-huh. You heard right. Not with a backhoe, not hiring anyone. By hand, with mattocks and shovels. This hole, by the way, needed to be 17 feet long, 6 feet deep and 6 feet wide. That doesn't sound so bad, does it? Well, it wasn't really, but we did encounter a couple of challenges. (We were also motivated by being told that it was impossible to dig such a hole by hand.)

First of all, we began digging this hole in August, and the weather was, predictably, hot and dry. The ground was correspondingly dry and hard. In addition, the first couple of feet down from the surface were a tangled mass of roots: salal, Oregon grape, thimbleberries, young fir and maple trees, and probably a few others. I noticed a funny thing: David somehow found he needed to be in Seattle frequently during the early part of the excavation, leaving me to dig most of this difficult top layer. Coincidence? Probably not, but no matter. The workout was great, and I discovered that I really liked using the mattock.

As the digging progressed, and the hole got deeper, we found we worked well in tandem, trading off every few minutes. I would descend with mattock in hand, and for a few minutes I'd whack away, busting up the hard rocky soil. Then I'd take a breather and get some water (it *was* hot) while David went in with a shovel and heaved out the loosened dirt. The deeper we went, the harder it was to fling the dirt out of the hole, so I was happy to have David do that part.

Somehow it seemed easier for me to swing the mattock, perhaps because I wasn't as strong in the upper body.

We worked on that project off and on for roughly two weeks. David especially liked bringing out the boom box and playing Broadway music: the soundtrack to *Oklahoma!* and the wonderful Cole Porter music of *Anything Goes* were our favorites to dig by. At last, the day came to install the tank. Terry, the very nice man who came to install it, stood at the edge of the hole and gazed at it a while, shaking his head gently. Finally he turned to us and said kindly, "You guys are crazy."

Terry later asked me to e-mail him a photo of the hole; apparently his colleagues didn't believe it when he told them about us digging the hole by hand.

Nonetheless, the job was done. Flush with confidence and a sense of well-being after the two-week marathon workout, I felt ready to start digging a root cellar. However, the reality of the approaching fall and cold weather prevailed. The house was mostly in good shape by that time, so we turned our attention to getting in a supply of firewood for the winter.

It's beautiful here, peaceful and quiet. Since our property is at the end of the road, and our house is about half a mile in from the gate, we have no traffic anywhere near us. Occasionally a bright orange Coast Guard helicopter (David calls them the Coasties) or some other aircraft flies over, but more often the only thing flying overhead is a wild bird. From every window, we see trees and grass and sky. It's lovely to walk down the road to open the gate, with the sun streaming through acres of trees. If you stop and listen, you'll hear the chatter of Canyon Creek far below. Or was that the wind in the trees? It doesn't matter. It's an amazing place, whether you call it a farm or something else. We own it, we love it, and every day when we get up, it's another beautiful day in paradise.

CHAPTER 3

Daydreams

I'D ALWAYS WANTED TO LIVE ON A FARM. I never understood this, since I grew up in Seattle; you'd think I would be a confirmed city girl. My twin sister and I were born in Champaign, Illinois, a university town in the heart of corn country. When we were about three months old, my father, who had just received another post-graduate degree from the University of Illinois (go Illini!), got a job as a computer engineer at Boeing, so we moved to Seattle. And although I lived there all my life until David and I moved to our farm, I had always felt more at home in small towns. Even when traveling, I tended to gravitate toward country settings. My mother grew up just north of Chicago, and used to tell stories of farm life and the pigs she raised. I remember feeling a twinge of envy now and then, wishing I had been able to grow up on a farm. But for various reasons, I had stayed in Seattle, most of those years right in the heart of the city.

David had dreamed of living on this property since he was a young boy. He was born in Indiana, and his family moved to the Seattle area in the mid-1950s, when David was five. He and his four sisters often spent part of their summer vacations here with their grandparents. David remembers loving it and hoping to live here someday. Eventually he made firm plans to move here as soon as he retired from driving city buses in Seattle. Long before that time arrived, he had discovered that no one else in his family was as interested in living here as he was; it was too far away from the city, too inconvenient a kind of life without electricity and had no neighbors nearby. So David made his plans, and continued working.

After his grandmother died in the early 1990s, David bought the property from the estate. Since he was still working in Seattle, the house was unoccupied for a time. He tried renting it out, but it was difficult to manage from a distance, and renters usually wanted to be there in the summer but not in the winter. Later he arranged for a tenant/caretaker, who stayed here for the last twelve years before David's retirement. In the meantime, David came up fairly often on weekends.

After I met David in 1999 (yes, we met on a city bus; that's a book in itself), he told me about his "country place," and brought me here occasionally to visit. I didn't even know what "off the grid" meant; all I remember noticing was the lack of electric lights. By the time we married in August 2000, with nearly six years to go before David was due to retire, I was not just looking forward to moving to the farm; I was doing some planning and dreaming of my own.

Meanwhile, David was working a lot of overtime; his pension would depend on how much he earned the last two years before his retirement. I appreciated that he was doing all he could to provide for our future. So although we had talked about it and I knew more or less what to expect, the reality was a little hard to take. He was coming home later in the day, so we had less time together. Also, he was understandably more tired; even when we were together at home, he often simply wanted to relax, look at his e-mail or whatever.

The last year before David retired was especially difficult for me. He was working six days a week and didn't use very much of the vacation time he was entitled to. I sometimes rode around on the bus with him, just to see him a little more. I have to admit that, about two years before he retired, I began to be concerned about the transition from this life — where we didn't have much meaningful time together — to a life where we'd not only be going through two transitions (his retirement and moving), we'd also be around each other much more. In fact, to me it really seemed like going from one extreme to the other. Somehow, knowing all this well ahead of time and thinking about it and planning for it didn't seem to make that much difference. I realized that you can't always anticipate everything, especially when it's a totally new experience, and we would no doubt make mistakes and have to learn as we went along. But still! The move being spread over a few months should have made things easier, but in the end, it was more stressful than we'd expected. By September, I was tired and so ready to be done with the moving process. And in my heart, I was definitely no longer in Seattle.

Ready to Grow

Certainly one of the things I took for granted about living at the farm was that I'd be doing a great deal of gardening. I'd always loved gardens, and even as a kid, I looked forward every spring to helping Mom with starting the seeds, digging, transplanting, watering, weeding and, of course, harvesting. As a young adult, for years before I was married, I hadn't done much gardening at all, except for a few potted plants on the little deck of my central Seattle apartment. Even after we were married, there wasn't any gardening to do other than mow the lawn, rake leaves and occasionally cut back the invasive blackberries and trim a few shrubs.

I do remember planting flower bulbs once or twice. Our backyard, although good sized, faced east and was largely surrounded by trees, so nearly all of it was in shade most of the day. We had a few pots on the east-facing deck, but somehow I never managed to grow much besides a few herb plants and bulbs. (The feral cats who hung around because our next-door neighbor fed them also tended to use our flower pots for their personal Sanikans, which didn't help the plants.) David always felt it wasn't worth putting much effort into gardening there anyway, since we already knew we would be moving in a few years. So I tried to be patient in those days, spending a lot of time thinking about all the plants I wanted to grow once we had the space and the time to do it.

You probably won't be surprised to know that, the first year at the farm, I pretty much went crazy with the gardens. I was dazzled by the relatively huge plots that were suddenly available to me. David and I both put in time with the rototiller that spring and early summer. I had gone a little overboard with the seed-buying, spending around $200. But virtually all of those seeds did get planted that year. We had talked at length about trying to raise a significant percentage of our food, and although I had spent a many hours planning and organizing, I did make some big blunders.

First of all, I was trying to do way too much the first year, considering I was also cleaning, painting and furnishing the house. Oh, and we were moving, too! Ugh. Needless to say, even though I worked hard at getting the seeds planted and was fairly consistent with watering, it wasn't very long before the weeds took over and I quickly lost control of the situation. I had planted the kitchen garden (about 1,400 square feet), the lower garden (about 3,000 square feet) and the potato patch (another 1,000 square feet) all at the same time. It was exciting, it was exhilarating, and I truly enjoyed being able to have a garden for the first time in years. Obviously I didn't account for the fact that I was completely out of practice and not very realistic about managing all that was going on. We did, however, harvest a lot of food that year in spite of all this.

The first frost of autumn (we didn't yet know when the cold season began or ended here) came sooner than expected, in early October. I ended up having a great amount of produce coming out of the gardens at once, including large harvests of cabbage and green tomatoes. I really had no idea what to do with it all, but I was determined to not let anything go to waste. I'd had years of experience with canning, and I planned to use some of the cabbage to make sauerkraut. But all those green tomatoes! I started looking through my canning books, hoping to find a recipe that used both cabbage and green tomatoes. And guess what, I found one! Piccalilli, something I'd heard of but didn't actually know what it was. I was pretty sure I'd never eaten it, and positive I'd never made it before.

Well, I made and canned a fair amount, and it was delicious. With lots of green tomatoes left, I put up quarts and quarts of green tomato pickles. I still have a couple of jars of that pickle; I like to add a jar to the pot when making chicken broth. But I didn't waste any of that cabbage or green tomatoes, so I felt good about that.

Growing vegetables was just one of the big ideas we'd had about becoming more self-sufficient. We weren't aiming to separate ourselves from society or anything; we really just hoped to become less dependent on grocery stores, fossil fuels and Big Ag in general. At that time, we hadn't talked about issues like doing this all organically. Terms that are commonplace now, like "sustainable," "green" and "locavore," were rarely heard even a few years ago. So in spite of all our talking and planning and daydreaming, we did many things in the early days without knowing if they would ultimately fit in with our long-term plans for the farm. And we hadn't even gotten to the point of raising animals yet.

CHAPTER 4

A Slippery Slope: Which Comes First, the Chickens or the Homework?

*W*HILE BROWSING AT THE LOCAL FEED STORE *for treats to tempt the wild birds visiting her backyard, she found herself captivated by the adorable little chicks peeping under the heat lamp. The next thing she remembered was all the noise (and what was that smell?) coming from the living room. Uh-oh, she thought. Now what do I do?*

Sound familiar? Well, that's not surprising. It's so easy to just bring home the cute little fluffballs and assume that it can't be all that hard to raise a few chicks. Frankly, it isn't especially hard, if you know what to do and how to do it. Ah yes, there's the rub. Chickens and other poultry (especially young birds) have specific needs: housing, feed, water, grit, bedding, warmth, even the company of other birds. And when they get bigger (they do, you know), what will you do then?

Not to put too fine a point on it, but have you actually thought about why you wanted the birds in the first place?

Chapter 6 will detail the planning process involved in preparing to raise poultry. But first, I want to explain why I think it's important to plan ahead before bringing home your little bundle of joy. Chances are that, once you get started, you'll be living with poultry for a good long while, so it's worth putting some time and effort into preparations.

When we first started talking about getting a few chickens, we were so excited. David remembers his grandmother having chickens here when he was young, but it was something entirely new for me.

We figured on producing enough eggs for ourselves, although we had no clue about how many eggs chickens laid, which breeds were best or anything at all of practical use. Eventually, we thought, we would breed the birds, let them hatch babies in the spring and slaughter extra roosters and older hens for meat once in a while. It all sounded easy on paper.

Note to self: Try to schedule planning meetings before Happy Hour, not after.

I remember my first perusals of the hatchery catalogs. Fascinating. Murray McMurray's catalog was especially interesting. It featured stylish painted illustrations of dozens of poultry breeds, along with each breed's conservation status: how the American Livestock Breeds Conservancy (ALBC; see Appendix B) ranked them in terms of rarity, based on population polls and numbers of breeders. Being a soft-hearted sentimental fool, I wanted to raise lots of the kinds that were listed as critically endangered. Like most well-meaning people, I thought I would be doing the world (or at least the poultry world) a service by helping to perpetuate a rare breed of chicken or turkey or whatever.

Of course, I didn't know the first thing about raising poultry then, much less anything about how complex a really good breeding program might be. Frankly, I was doing what I usually abhor: actually contemplating rushing into something without adequate research, basing an idea on emotion rather than reason, without even considering our actual short- and long-term plans. I've learned over time that plans and goals tend to change as you go along; you write down ideas, you make plans and lists, you think things out (or imagine you do), and eventually you pick a place to begin and then do it. Looking back, it's amazing to me how much our lives and what we're doing on the farm have changed in just a few years.

We did talk ourselves into getting several critically endangered breeds: Houdans (a rare crested old French breed of chicken), Nankins (a tiny true bantam that we were mainly interested in because of their reputation for being broody) and Midget White turkeys, which we liked mostly for their smaller size and reputedly calm temperament.

The Houdans turned out to be a mistake. Although they are described as a good meat bird, none of ours ever got anywhere near the size the catalog claimed they would. The real problem, though, wasn't obvious until we had had them for a couple of years; for the first time, we started losing birds to bobcats that year.

One by one, the poor Houdans disappeared, and we finally realized why: their large floppy crests fall in their eyes, impairing their ability to see anything that's not more or less under their noses. We could walk right up to them, and they would simply stand there, looking at the ground in front of them. Usually they didn't move until one of us actually tried to pick them up. All the other birds get out of the way when you walk toward them. So eventually we lost nine of the ten Houdans we had started with, and we felt terrible about it. The remaining hen I gave to my sister to add to her flock. It seemed to do fine there, and even started laying eggs again. This was another letdown with the Houdans. They did lay eggs at first but were never consistently productive. (We knew this because they were the only hens in our flock that produced white eggs.) We eventually

Nankin hen on nest.

figured out that birds tend to slow down or stop laying when they're under lots of stress, such as when predator attacks have occurred. This was one of many lessons we only learned over time — and painful experience.

The Nankins, on the other hand, have done extremely well here. We like them for many reasons. According to what I have read, they're not supposed to be a very cold-hardy breed, which was the only misgiving I had about trying to raise them here. We typically have several days in the teens and twenties during the winter, and a week or so of single digit temperatures. However, the Nankins have been housed in an uninsulated mobile coop through several winters now, and we've had no apparent cold-related issues such as frostbitten combs. (More on cold-weather issues in Chapter 31.)

Nankins are tiny birds. Our Nankin roosters top out around thirty ounces; some of the hens reach barely a pound. But they more than make up for size in attitude. We've seen the Nankin roosters stand up confidently to large New Hampshire roosters, the wild Canada geese that sometimes stray into our yard looking for food and even the turkeys. They are fiercely protective of their flock and seemingly fearless. However, Nankin roosters rarely fight among themselves and have gentle dispositions. They like to be picked up, and a few of the hens even habitually wait on top of their coop at night for one of us to pick them up and put them into their coop by hand. (All together now: Awww ....)

Something else in the hatchery catalog that intrigued me was information about the Cornish Cross meat birds (broilers): five to six pounds carcass weight in six or seven weeks, so the description claimed. David and I had previously discussed raising a couple of batches of meat birds every year, but we didn't know the first thing about sustainability or other reasons why fast-growing hybrids might not be what we wanted. I was also curious about the fact that, of all the many chicken breeds in the catalog, only the Cornish Cross males cost more than the females. I gather this is because the males grow out to slaughter weight a week or two faster than the females, which I guess matters if you're raising them to be sold. In any case, I don't

Nankin rooster crowing.

really remember now why we then didn't go ahead with our plan to raise meat birds. Possibly we were realizing there was much more to learn, and it seemed prudent to go one step at a time, adding new ventures as we gained experience and refined our goals.

It turned out that this was an excellent strategy. As it happened, we ended up keeping chickens and ducks mainly for laying. For all our planning, so much turned out differently from what we had envisioned. We couldn't have known that, within the first two years, we would be selling chicken and duck eggs to a local restaurant. Or that we would ever have more than a hundred and fifty birds at one time.

True, it's been a grand adventure. But believe me, it's a slippery slope.

CHAPTER 5

A Rooster Called Charlemagne

I T WAS THE DAY AFTER CHICKEN DAY. Around 5:30 in the morning, we heard crowing. The coop being about 65 yards from the house, I was surprised at how loud it was. "There he goes," mumbled David. He turned over and pulled a pillow firmly over his head.

As soon as it was fully light, we went down to the chicken yard. The chicken feed was stored in the coop, and David was lugging a three-gallon poultry drinker. Although we planned to let the birds free-range, we felt that it would be best to confine them in a large fenced area for a few days. We reasoned that this would give them time to make the transition to an unfamiliar place (with unfamiliar owners), while keeping them relatively safe.

I stood by with my camera as David unlatched and opened the front door of the coop. We stayed off to one side, waiting. Soon the rooster, Charlemagne, leaned his head out the door, looked around suspiciously and stepped cautiously out onto the ramp. After a minute or two, he apparently noticed that nothing had, in fact, attacked him. He walked quickly down the ramp, followed closely by the six hens.

In the excitement and general hilarity of the photo session that followed the appearance of our first chickens, we were blissfully clueless that we were about to find out how much we had yet to learn.

You're probably wondering why I chose a name like Charlemagne for the rooster. Although in general I haven't the slightest inclination to name our animals, I had planned all along to name our first rooster Charlemagne after the infamous rooster in Peter Mayle's *A Year in Provence*. You know, "le bloody cock" who crowed all night, infuriating the Parisian neighbor, Madame Hermonville, and her visiting city friends.

Luckily, our nearest neighbors are two miles down the hill, and as far as I know, they're not Parisians. However, like the real Charlemagne (an empire-building Frankish king who eventually conquered much of Europe), our rooster was aggressive, protective of his harem and most definitely not monogamous. When he started sneaking up behind us and attacking us without provocation, well, his little empire came to a premature — and yet not untimely — end. You know, le bloody stockpot.

Oh, by the way, we *did* get one egg from our new hens the first day.

Charlemagne.

CHAPTER 6

Poultry from Scratch

BEFORE WE HAD OUR OWN CHICKENS, I honestly didn't know the difference between a fryer and a roaster. I knew even less about turkeys. Ducks? Are you kidding? Having humbly accepted my complete ignorance on the subject of poultry, I decided to educate myself.

About six months before we moved from Seattle to the farm following David's retirement, I subscribed to *Backyard Poultry*, a fairly new magazine serving small-flock poultry enthusiasts. My very first issue had a brief article about choosing the "right" chicken. Excellent! I thought, just what I need. It included a handy chart comparing certain attributes of a dozen or so chicken breeds.

Right away I encountered unfamiliar terms and began to feel even more ignorant. ("Broody"? What the heck is "broody"?) The chart compared things like egg color, mature weight of the bird, personality, cold-hardiness and whether the breed was better suited to a free-range situation or confinement. While it was informative as far as comparing those particular characteristics of those particular breeds, I soon realized that it really didn't help me to figure out what breed or breeds would actually work well for us, in our particular situation.

For example, there are just the two of us. If all I want is a few eggs every day, how many hens do I need? We live in the foothills of the Olympic Mountains. Even if I choose a breed described as cold-hardy, will I need to insulate or even heat the coop in the winter? How much time can I expect it to take every day to care for, say, twelve chickens? And by the way, what *does* "broody" mean? The term was never defined in the article; I daresay the writer took for granted some basic knowledge on the part of the reader.

In between the bi-monthly issues of *Backyard Poultry*, I got books from the library and spent some time searching the Internet. I was hoping to find not only further general information about poultry husbandry, but also answers to my specific questions. I began seeing and hearing more and more stories about people who ordered a batch of chicks or bought them from the local feed store and only then started to figure out what in the world to do with them. Being determined to plan ahead, especially on a subject I knew so little about, I continued to read and surf. I concluded — with some discouragement — that there were some serious gaps in the information available, and I still had many unanswered questions.

In addition to the questions already mentioned, others occurred to me that seemed to be important in the long run. Given everything else we expected to occupy ourselves with on the farm, how much time could we realistically spend managing chickens? Might turkeys and/or ducks work better for us, and could one or both of them peacefully co-exist with chickens? What issues might we encounter if we decided to raise the birds organically? What's the difference between purebred and hybrid types? What potential predators will we have to guard against? How might our farm being off the grid affect our experience?

After a lot of work, I eventually came up with a list of what I thought I should look for in a chicken breed. If the breed didn't meet these criteria, it wouldn't make the short list. (These are not in any particular order.) First, **friendliness**. Besides our own daily interaction with the chickens, we expected our young nieces and nephews to visit the farm, and we wanted chickens that like being around people. Second, **cold-hardiness**. Chickens are relatively small animals, and I really had no idea how well they would do when the temperatures dropped below freezing. Third, **good foraging abilities**. Since we were hoping to free-range our flock, we wanted only birds that could provide some of their own food. Fourth, the breed should be **dual-purpose**, good for both egg and meat production. Fifth, tendency toward **broodiness**. Once I figured out that this meant the hens are inclined to incubate and hatch eggs, I thought it was a good thing. Finally, given everything else I wanted in a chicken, it made sense to look at purebred (heritage-breed) birds, rather than production-type hybrids.

You might be wondering why I spent all that time on research, or even how in the world I came up with all those questions. After all, they're, well, chickens. True, but although we started from a place of knowing nothing about raising poultry, we did know that whatever birds we ended up with would be considered partners in the process of making our farm — and lives — more sustainable. These are living things, and we feel we should put our best effort into taking care of them.

The "Poultry from Scratch" concept shows the reader how to thoughtfully plan for success with heritage poultry. It is a simple worksheet (see Appendix A) designed to clarify the planning process by raising a number of questions that should be considered before

Four-week-old chicks are already enthusiastic foragers.

making decisions. You may already have thought of some of these points; others may come as a surprise to you. The worksheet also helps you to answer these questions. For example:

- How many hens do I need to supply eggs for a three-person family?
- How might the climate in my area affect my choices?
- What kind of poultry housing is best for my situation?
- How many chickens can I raise in a small backyard?
- Are heritage breeds really the best choice for me?

I learned a few important things through my own process, which led me to develop this worksheet. First, I didn't know all the right questions to ask. Second, even after all the reading I had done, I still had some unanswered questions, most often about things specific to our situation. I believe you will find the Poultry from Scratch worksheet very helpful and easy to use.

CHAPTER 7

Turkeys Are People Too —
but They're Not Chickens

"**W**HY WOULD YOU WANT TO RAISE TURKEYS? They're so stupid!" Sigh. We got pretty tired of hearing that one whenever we mentioned that we were going to start raising turkeys. Oh, and how I wish I had a nickel for every time someone asked, "Is it true that turkeys are so stupid that they'll look up during a rainstorm and drown?"

Of course, being the kind of person who has to see everything for herself (and not happening to believe everything I read on the Internet), this made me even more interested in turkeys. About a year after getting our first chickens, we bought some day-old turkeys: ten Midget Whites and fifteen Narragansetts.

According to what we had read, the Narragansetts would get quite a bit bigger than the Midget Whites. We figured most of our turkeys would be slaughtered for family and friends around the holiday season, and we wanted to be able to offer them a choice of sizes. Interestingly, we learned later that year, from our chef friend Gabriel, that small turkeys are the best. "If you want twenty pounds of turkey," he advised, "get two ten-pounders."

Since buying those first turkeys in the spring of 2008, we've come a long way. We knew very little then about raising turkeys, in part because we found few sources of useful information. In fact, although more resources are available now than just a few years ago, my main reference continues to be a book first published in 1929 called *Turkey Production* (see Appendix B).

In studying *Turkey Production*, at first I was confused by the absence of terms like "heritage" and "organic." Eventually I realized that was because the faster-growing hybrid turkeys common today had not yet been developed for commercial production. Turkey operations were organic, and all the turkeys were purebred — what we now commonly refer to as heritage breeds. It was very interesting to read all this when I had almost no practical knowledge or experience with turkeys; we were, after all, only in our first year of raising chickens.

In the 1940s, development of the Broad Breasted turkey was proceeding rapidly. Somewhere along the line, breeders decided to select not only for faster maturity, but also for ever-increasing amounts of breast meat. I've seen a few adult Broad Breasted turkeys (live ones, I mean), and honestly, to me it was a pathetic sight. I remember one tom that could barely walk; he was so front-heavy with breast meat that he literally tipped forward when he walked. Another astounding fact we learned was that Broad Breasted turkeys, by virtue of their great size (Giant White tom turkeys can reach 45 pounds or more) and huge breasts, cannot mate naturally — they must be artificially inseminated.

All in all, we concluded that, although Broad Breasted turkeys are cheaper to buy as day-olds, heritage turkeys, which cost about twice the price, were the right choice for us.

One feature we like about heritage turkeys is their ability — and inclination — to forage. Like our other birds, turkeys tend to head straight for the feeders when we let them out in the morning. After a quick breakfast of organic grain mash, they disperse for a leisurely day of foraging on pasture. I've noticed that turkeys, compared with our chickens and ducks, are especially fond of clover. They also love apples. We love them for cleaning up the windfall apples from the old trees David's grandmother planted years ago; it really helps keep the deer out of the yard.

I should emphasize that the recent trendiness of heritage turkeys was not a factor in our decision to raise Midget Whites and Narragansetts. The more we learned about the inherent problems in

Midget White turkey flock foraging.

hybrid-poultry breeding systems, most especially the suffering of the birds that frequently results from over-selecting for single traits at the expense of overall health, the more the right choice became clear. We believe that purebred animals, capable of reproducing true to type, are simply more sustainable than hybrids. Yes, purebreds usually mature more slowly, increasing the cost of raising them. On the other hand, if we were buying new turkeys every year instead of breeding them, I'd bet the numbers would even out before long.

In any case, since we aren't trying to make a living raising and selling cheap turkeys, our main consideration does not involve fast growth or enormous quantities of breast meat. The relatively few turkeys we sell each year, though, are super-premium and organically raised. Our customers love to come to the farm, where they see our birds free-ranging; it is obvious to visitors that we really do care about the birds' quality of life, not just about how much money we might make selling them.

I mentioned that we raised two kinds of turkeys that first year. It really didn't occur to us at first that having two breeds would be an issue. I suppose we didn't really know then if we were going to continue raising turkeys beyond that one year; most often, when we've tried something new, it was with the understanding that we would see what happened. I still think this is a good approach, for small operations like ours: start small, add new things a little at a time, over a period of time, and you'll figure out what works for you in your particular situation.

With turkeys, for example, we honestly didn't know what kind of predator issues we might encounter. Even small turkeys like Midget Whites are good-sized birds. We speculated that a bald eagle might be able to take down a Midget White hen; this was partly based on our assumption that, as our birds were enclosed at night, the only daytime predators we had to worry about were hawks and eagles. Actually, the few turkeys we've lost to predators have mostly been killed by cougars. We found one young tom dead in an enclosed roost area and suspect it was killed by a raccoon. The only young turkeys we lost were a few four-week-olds killed by a weasel that got into their coop at night.

Aside from predator issues, we discovered other differences between chickens and turkeys as we went along. Again, the lack of available resources forced us to learn by experience. We've talked to a number of people who have picked up a couple of baby turkeys at the local feed store and simply raised them with their chickens. This is not necessarily wrong, but it is important to keep in mind some basic differences between turkeys and chickens.

First, turkeys need much more protein in their diet than chickens do in order to grow well. We're fortunate that our feed supplier offers three different turkey mash formulas: 28 percent and 24 percent protein starter rations and a 21 percent grower mash. I say we're fortunate because, depending on your location, turkey feed can be challenging to find. An acceptable substitute is a game bird feed; it can vary in protein content but usually has enough protein (25 percent or more) to support good growth in turkeys. However, an actual turkey ration is preferable as it is specifically formulated for their nutritional needs. Turkeys are, after all, much larger than most game birds.

Being large, turkeys will eventually need more coop space per bird than chickens. They also need larger, stronger roosts. The first turkey house I built is six feet by six feet inside, with a walk-in door at the side and another door at the front. The roosts we use for Midget White turkeys are cedar two-by-fours mounted with the wide (3½") side facing up. Believe me, even small turkeys have large feet! Our oldest tom, who is five years old, weighs over twenty pounds; other heritage breeds can grow to over thirty pounds. In our turkey house, we've found that a six-foot two-by-four can easily support three or four adult Midget White turkeys. If the roosts are any longer, they will probably need to be supported in the middle, especially if you raise one of the larger heritage turkey breeds.

Incidentally, some claim that Broad Breasted turkeys do not need roosts. Presumably this is because by the time they're big enough to want to roost, they're too heavy to get up on one. Turkeys do love to roost, though. I suggest you provide low roosts for Broad Breasted turkeys, so they can roost if they want to but avoid injuring themselves on higher roosts. Even if they are too heavy to fly up onto a roost, their instinct is to roost, so they may attempt this. It's better to be prepared than to risk your birds injuring themselves by trying to roost on something that isn't strong enough to support their weight.

This reminds me: Although turkeys, especially the smaller heritage types, are very good fliers, it's important to not place their roosts too high up. Our turkey coop has three roosts. The first is barely fifteen inches from the floor, the third and highest about twenty-four inches. As turkeys get bigger, they can be susceptible to leg injuries if they come down too quickly from a height, so make sure the roosts go gradually from lower to higher. This enables them to hop easily and safely from roost to roost.

Turkeys are very curious and sociable birds. They seem to want to be around people and are inquisitive about what we're up to. If we're in the house, they will walk around the house, looking in every window until they discover where we are. They love to hop up on lawn chairs or whatever is handy and then sit staring at us through a window. If it wasn't so darn funny, it might be a just a little bit creepy!

As I mentioned, turkeys do love to fly up and roost on things. Like the roof of our woodshed. Like our cars. When we have overnight guests, we often spread an old moving blanket over the hood

Our very own peeping Toms.

of their car to prevent the turkeys from scratching the paint. We've often been asked why we don't simply clip their wings so they can't fly. We debated this and decided that they were safer being able to fly. Midget Whites are very good fliers, and we feel that it's preferable for them to have the option of flying into a tree when they are trying to avoid a predator. My sister, who raised a few heritage turkeys last year, told me that clipping their wings didn't prevent them from flying high enough to get over a fence or roost on the roof of a shed. If you are going to clip wings on any poultry, you also have to remember to do it every year, since they regrow new feathers after their annual moult in the fall.

Toward the end of our first year of raising turkeys, we decided we really wanted to keep going with it. We figured out, rather belatedly, that if we were going to breed turkeys, it would be much easier with only one breed. Since all our birds free-range during the day and share pasture space, obviously we would end up with some mixed-breed turkeys if we didn't separate them during the breeding season. We opted to stick with the Midget Whites for several reasons. Being smaller birds, coop and brooding space was easier to deal with.

Young turkeys on a ladder.

Naturally, they eat less than larger turkeys, and they are excellent for-agers. Also, we found that the Narragansett toms especially tended to frequently fight amongst themselves, something the Midget Whites rarely do. Finally, to our surprise, many of our customers, although they were initially skeptical about the small size of our turkeys, found that they actually had plenty of meat to go around. Plus, whether roasted or smoked, they are truly, truly delicious. (See Chapter 26 for more on cooking heritage turkeys.)

Here are a few of the major differences between turkeys and chickens:

- Turkeys don't lay eggs year-round. The laying season of heritage turkeys is generally between March and June or July. There are exceptions to every rule, though; one year, a Midget White hen hatched a clutch of eggs on New Year's Day!
- Turkeys incubate their eggs for 28 days, compared to 21 days for chickens.
- Heritage turkeys can reach a good slaughter size in six months. However, we prefer to raise them to seven or even eight months, as we think the quality of the meat is better.
- Turkeys are more sociable than chickens and prefer to hang out in groups.
- Given a choice, turkeys will roost outdoors at night, usually up in a tree.
- Our chickens all go into their coops on their own every night. Our turkeys hang around outside, waiting for us to escort them to their coop. This is not because they're too stupid to know where to go, it is simply their routine.

Turkeys, as we've come to know and appreciate, are not just like large chickens. They have their own personalities, and I think they're a good deal friendlier than most chickens I've met. Although we have sometimes felt ambivalent about continuing to raise turkeys, overall, it's been a genuinely enjoyable experience.

CHAPTER 8

Everybody Look Busy — Here Come the Ducks!

> Ducks are the perfect backyard livestock because they
> are so well-behaved.
> — Eliot Coleman and Barbara Damrosch,
> *Four Season Harvest,* Chelsea Green, 1999

"OH, BY THE WAY," David said casually to me one fine April morning. "I ordered some ducklings."

"Uh-huh?" I replied. We were in the middle of morning poultry chores, and I wasn't really paying close attention. After a moment, I realized that David was standing still, waiting for a reaction. The light dawned.

"You mean *live* ducklings?" I was trying very hard to sound casual myself. I have to admit that I was somewhat shocked, having only a hazy memory of talking once, in passing, about someday getting ducks. Suddenly, it appeared, ducks were happening.

I picked up the baby turks I had inadvertently dropped and took a deep breath. "And when is the blessed event scheduled to occur?" I asked. Please don't say it's happening tomorrow, I prayed. David mentioned a hatch date about three weeks down the road, and suggested anxiously that I might want to start breathing again.

"Well," I said, after regaining consciousness, "you'd better bring home some books from the library. I don't know the first thing about raising ducks."

Did I mention that I like to ask questions first? Since the little darlings were already ordered and presumably the eggs were in the incubator (not that I knew how long duck eggs had to be incubated), a slight nervousness was creeping up on me. By the next day, I admit, I was feeling fairly panicky.

When, I asked myself, was I going to find time to build yet another coop? For that matter, what kind of house do ducks prefer? Beyond having a vague impression that ducks don't roost like chickens, I had no idea what their housing needs might be. What do ducks even eat? I had not a clue.

Although David had no luck finding a library book about raising ducks, he did eventually find a decent one at a thrift store in Seattle (see Appendix B). It was apparent that, as with turkeys, few helpful resources on ducks were available. I learned pretty quickly that ducks are quite different from turkeys and chickens. I know that sounds obvious, but bear with me.

It's important to know that ducks tend to be fairly high-strung, compared with the much more mellow turkeys. Ducks hate stray light in their coops at night; it may even make them panic. The book David found, by Rick and Gail Luttmann, turned out to be a valuable source of information. Originally published in the 1970s, it was written for the beginning small-scale waterfowl enthusiast. Leafing through, I found a good example of a mobile duck coop and proceeded to build.

This first duck house is six feet long, three feet wide and about three feet high at the front of the sloping roof. I sized it for the eight ducks we were expecting. The book said that one and a half to two square feet per adult bird was optimum for laying ducks. The coop is on wheels and is fairly easy to move, even when it is full of ducks. We do move it around from time to time; for example, I like to move it into the shade of the old willow tree during the hot part of the summer. We also sometimes reposition it to be out of the wind if we're expecting a severe windstorm.

One thing I would definitely do differently when building duck coops is to have solid wooden floors. Several of the poultry coops I built that first year had half-inch mesh hardware cloth floors: I assumed that this would make the coops easier to clean. However, I have concluded that, in most cases, wooden floors actually work better. Ducks are especially messy (she understated). They drink copious quantities of water, dabble in mud puddles all day long, and their poop is naturally very wet — and abundant. Of all our coops, theirs definitely require the most frequent cleaning. Although I usually toss in a layer of fresh bedding every other day, in an attempt to keep up with the moisture level between cleanings, we clean out the coops once a week or so. I wouldn't wait much longer, if I were you. You don't want to be impressed by how heavy that bedding is when you rake it out. This is especially true in the winter, when the birds spend more hours in their coops than during the long dry days of summer.

Incidentally, if I could give just one piece of advice about coop design in general, it would be to make them easy to clean. If you can put a removable door at the level of the coop floor somewhere, and make it wide enough for a steel rake to easily fit in the opening, trust me, you'll be glad you did. It's lovely to be able to open the door, rake out all the bedding onto a heavy tarp and haul it away to be composted. When you clean out coops frequently, as with ducks, it really makes sense to do all you can to make it a relatively simple chore.

Mobile coop.

One issue we're trying to figure out with the duck coops is what to do about nest boxes. Based on my book's recommendation, I put a nest box in one end of the coop, raised a little off the floor. The ducks did use it, but they also spent the night in it, having a great time kicking out all that nice dry bedding and filling the box with poop. They also laid eggs all over the floor of the coop. Still do, for that matter.

We have noticed that our ducks usually like to pick one corner of the coop to lay eggs in, although this depends on how much space there is and how many sleep in the coop. Ducks usually lay their eggs during the night or early in the morning, so by the time we let them out, almost all their eggs will already be there. A few seem to prefer to find an outdoor nest to lay in, and some days it's felt like one big Easter egg hunt around here. However, they tend to go back to that same spot every day, so if you find it, you'll know where to check the next day.

Blue Swedish drake.

The ducklings David ordered were of two varieties, Blue Swedish and Khaki Campbell. Yes, we had to learn the same lesson with ducks as with turkeys about the relative merits of having more than one breed at a time. In our defense, I can only say that we had to start somewhere. I mean, how would we know, before actually trying it, what breeds would work best for us? It was good to be able to compare egg-laying ability, personality, foraging ability and so on. We had no way of knowing that, within a year of getting our first ducks, we would discover a niche market for duck eggs in our area. This has completely altered our original ideas about why we might raise ducks at all. I tell you, once people found out that we had organically raised duck eggs, the demand has grown every year, and consistently outruns the supply.

By the way, of that first batch, we still have two Blue Swedish ducks; David calls them the Swedish Bikini Team. As I write this, in early 2013, they are going on five years old. They are beautiful, relatively calm for ducks, and although their egg-laying has slowed down over the years, they produced a respectable number of eggs last year.

Young Khaki Campbell ducks.

This is not a scientific assessment, but I think ducks are the smartest of the three kinds of birds we raise. I read recently that ducks are often used to train some kinds of herding dogs because they tend to stay together as a group and are consequently easy to move around. This is easy to observe when we escort the ducks into their coops at night. Take a step to the right, and the whole group of them moves to the left, and vice versa. I suspect this grouping behavior also accounts for the comparatively few predator losses our ducks have suffered. According to most of my sources, ducks are supposedly more predator prone than chickens or turkeys. This, to me, is a good example of how everyone's situation is different. You really can't take it for granted that what works for one person is going to be ideal for you, or that you'll encounter the same problems as someone else. It would certainly be convenient if we could generalize that way, but my experience has been that many factors are likely to be unique to your situation, and you will simply have to figure them out as you go along.

David likes to tell people that the main reason we have ducks is for slug control. It's true that I haven't seen a slug anywhere near my kitchen garden since the first season we had ducks. They love slugs! Chickens will sometimes eat slugs, although I have only seen them eat very tiny ones. I remember that first spring, as I was loosening the soil in my kitchen garden beds, the ducks would line up, single file (I swear this is true) outside the garden fence, waiting for me to toss slugs over to them. Occasionally a duck would catch the slug in mid-air and, with head quickly tilted back, gulp it down in one fluid motion. Excuse me, but I mean, ugh.

The ducks are prolific layers. They also seem to me to have a shorter off-season than the chickens do, laying eggs for about ten months before slowing down in winter. In the spring, when egg production is heaviest, we sell whatever eggs the Alder Wood Bistro doesn't need at several local retail stores. Khaki Campbells, generally reputed to be the champion egg-layers of the duck world, can easily lay over 300 eggs per year. Considering that they also moult in the fall and their egg production slows down, this is pretty amazing.

Chapter 9

Violet, Bumptious and Hampty

Bumptious: Presumptuously, obtusely, and often noisily self-assertive.

— Merriam-Webster Dictionary

Wいth the exception of our first rooster, Charlemagne, and our old tom turkey (cleverly named Old Tom), we have rarely named any of our birds. It's a good thing, too, considering how many we have now! Anyway, over the years, we have made an occasional exception to this rule. Usually this was a bird that had become something of a farm mascot. The first of these mascots was Violet, a.k.a. Violent.

Violet was a Buff Orpington hen, one of the original seven chickens we started out with. A few months after we got her, she started showing signs of not being well: she was keeping away from the other chickens and tended to hang out in a corner of the yard with her head hunched down into her shoulders. She didn't seem to be eating much, and was clearly losing weight. David started calling her Shrinking Violet, or Violet, for short.

We were concerned, of course, but we had no idea what was wrong or what to do. We called a local veterinarian who was nice enough to give us some advice over the phone. She said the most important thing to remember with sick birds is to keep them warm. It was late fall, and nighttime temperatures were consistently below freezing. We decided to bring her inside for a day or two, where she could stay warm by the wood stove and we could keep a closer eye on her. She ate a little bit and seemed like she was feeling better.

At that point, we moved her to a temporary outdoor pen, thinking it would be a good idea to isolate her from the flock, just in case she had some kind of contagious disease. She promptly escaped (we never did figure out how) and rejoined her flock-mates. This happened twice more. Each time she went back to the other birds, she seemed to get more aggressive and assertive about claiming her place in the pecking order; hence the modification of her name to Violent. Then she would start to go downhill again, and we would isolate her again. Finally, after the third round of this routine, she died.

We were glad that she wasn't suffering any longer, but we also wanted to try to learn something about what had happened, if possible. David did a post-mortem examination and discovered an undigested piece of raw potato in her stomach. We know now that raw potatoes are poisonous to birds (as well as pigs and some other animals). From then on, we have been careful to keep the birds away from anywhere we are growing potatoes, and any potato peelings to be composted are cooked first, just to be safe.

So Violet was our first real mascot. It probably should have been our rooster, Charlemagne, but he turned out to be an aggressive, mean-spirited despot, no doubt much like his namesake. He didn't live long enough to be a mascot, anyway.

Bumptious was one of Charlemagne's sons, hatched in the second batch of chicks he sired that summer. Actually, his first nickname was Little Shit. He was a friendly bird who liked to hop up on our laps when we were sitting outside. One time he jumped up on David's lap, whereupon he promptly squatted and deposited an impressive pile of poop on his shoe. David, who is much better than me at coming up with names for just about anything, chose the marvelous word "Bumptious," partly as being a more acceptable name to use in mixed company. But actually, the name fitted his personality really well (the rooster's, I mean, not David's).

One of the most difficult things we had to do that first year was to slaughter Bumptious. He had become progressively more like his father, aggressive to the point of sneaking up behind us and kicking at our legs with his spurs. That hurts, believe me, when a rooster's spurs make contact with an unsuspecting shin. Poor David was just a wreck that day. He was quite fond of Bumptious, and it was a terribly difficult decision to slaughter him, in spite of the rooster's behavior. As far as we were concerned, giving him away was out of the question. We could not, in good conscience, have given him to someone else, knowing how mean and aggressive he was. That was a tough day for both of us, but especially for David.

Hampty was our first New Hampshire rooster. He was a large beautiful, friendly bird who helped us decide to go with New

Hampty.

Old Tom, the patriarch of our Midget White turkey flock.

Hampshires as our choice of dual-purpose chicken. We had him for over two years, and we just loved him. He was calm, quieter than most roosters, very protective of his hens and never even a little bit aggressive toward us. We lost him in 2010, after he was injured by a bobcat. He survived the initial attack, but sadly he died a few days later. He had produced some beautiful chicks, though.

The only other real mascot we've had is Old Tom, the patriarch of our Midget White turkey flock. He was hatched in the spring of 2008. He is truly a sweet, sweet bird. He loves people, loves to be petted, simply adores being in the center of a crowd of people, especially if one of them is clicking a camera in his direction. He is also an excellent breeder, having fathered dozens of baby turks over the years. We honestly can't imagine ever slaughtering him, unless he were injured or ill to the point where we felt we had to end his suffering. Most likely, hopefully, he will live to a ripe old age here on the farm, enjoying his golden years being admired and photographed and petted by children.

We have many good memories connected with Old Tom, like all of our previous farm mascots, and it's hard to imagine life here without him.

CHAPTER 10

Heritage Turkeys (and Chickens) Are More Sustainable, and They Have More Fun

OVER THE PAST FEW YEARS, heritage turkeys have spent some well-deserved time in the spotlight. A Internet search shortly before last Thanksgiving revealed an impressive number of growers offering them for sale. If you were one of the many who shelled out a few extra bucks for one of these popular birds, what made you choose a heritage breed? Novelty? The trendiness of it all? Before you prepared and ate it, did you really know what makes a heritage breed different or better than the normal frozen supermarket variety?

I know I've said this before, but I think it bears repeating, especially if you're interested in raising turkeys and want to make a thoughtful choice. A "heritage" breed is a purebred, or standard, bird, while the usual commercial turkey is the faster-growing Broad Breasted hybrid. Heritage breeds are also defined as breeds that are "naturally mating" types. Yes, you heard right: Heritage-breed turkeys mate naturally; hybrids, mostly owing to their unnaturally large breasts, cannot mate. They must be artificially inseminated.

As David noted wryly, someone has to do it, but how would you like that job on your résumé?

Apparently, in someone's estimation, the minor inconvenience of artificial insemination is offset by the faster growth that brings them to your grocer's freezer two to three months faster than the heritage breeds. In addition, without artificial insemination, turkeys would

Mom turkey with new hatchlings.

not be available year-round. And if you do the math, you'll discover that, when turkey eggs are hatched in mid- to late spring, the babies are reaching slaughter size right around late November. I've never heard anyone else speculate about this, but I certainly have wondered if that's the real reason turkeys are so closely connected with the Thanksgiving holiday.

What if you don't want to raise turkeys yourself, but you do want to be a more responsible and informed consumer? If you are serious about learning more about where your food comes from, you have probably already started to think beyond the garden or greenhouse. See if you can find someone who raises poultry — it's really not hard these days — and start asking questions. If you live in a big city, ask at your local grocery or co-op. If your store doesn't offer heritage turkeys, they might be willing to order one for you. (Chapter 30 offers more ideas about connecting with sources of heritage poultry.)

Midget White turkey family out for a stroll.

By the way, "heritage" applies just as much to chickens as to turkeys (as well as other types of livestock). Many of these historically important breeds are now endangered. Heritage-breed meat chickens in particular have had a tough time since the ridiculously fast-growing Cornish-Plymouth Rock hybrid was developed in the 1950s.

There are just as many good reasons to buy and enjoy heritage chicken ... if you can find it. If you're like most people, you eat chicken more frequently than turkey. I heard that, later in her life, Julia Child regularly bemoaned the disappearance of the stewing hen from the grocery store meat section. The poultry industry favors the male Cornish Cross chickens because of their faster maturity, although the difference is only a week or so between males and females. Knowing what I know now, I think it's sad that the profit margin has become such an issue that slaughtering a chicken at six weeks is so different from slaughtering one at seven weeks.

I also suspect that the 'Rock Cornish Game Hen' is an idea someone came up with to deal with surplus Cornish Cross chicks at the hatcheries. (The "Rock" comes from the Plymouth Rock chicken, which is crossed with the Cornish chicken to produce the Cornish Cross.) And a clever marketing solution it was: Slaughter the pullets at two pounds (live weight), change the name of the birds a little bit, and hey presto, we now have an exotic-sounding miniature chicken that can be sold for much more per pound than their male siblings. It reminds me of the equally clever story behind the portobello mushroom. But don't get me started on that one.

CHAPTER 11

"The Mind of a Turkey"

"**H**EY, I JUST THOUGHT OF A GREAT TITLE FOR A BOOK**," David called from the kitchen yesterday afternoon.

"Oh? And what title is that, my love?" I called back after a moment. I was relaxing by the living room wood stove with a *New York Times* Sunday crossword puzzle.

"*The Mind of a Turkey*," he replied, walking into the living room.

"Well," I retorted, "That ought to be a pretty short book."

Okay, so that wasn't a very charitable remark. We have heard many fairly uncharitable remarks about turkeys, usually from people who have never actually raised them. Several years ago, when we were first contemplating the addition of turkeys to the farm, we heard this one frequently: "Why would you want to raise turkeys? They're so *stupid!*" This was often followed by the recital of one or more supposedly funny myths about turkeys. Yawn.

Since you've apparently read this far, you know we love our turkeys. You also know that, after raising both Midget White and Narragansett turkeys the first year, we settled on the Midget White as the right breed for us. As I write this, it's November, and Thanksgiving is just a few weeks away. So, being in the sentimental mood that the approaching holiday season sometimes brings out in me, I thought I'd share some of our observations about, well, the mind of a turkey.

First, we don't consider turkeys to be any less intelligent than chickens. Stop snickering! Turkeys are people too, and they have feelings. Seriously, we do think turkeys have more personality than chickens, and their mannerisms are fairly adorable, at least sometimes. We especially love their habit of following us around; they seem to just be insatiably curious about what we're up to. Anyone who's visited the farm knows that the turkeys appear en masse to greet newcomers. (We had a funny incident last year with a substitute UPS driver, who positively leaped back into his truck when the turkeys approached; the poor guy thought he was being attacked.)

At this time of year, the turkeys that were hatched in the spring are mostly between six and seven months old. The young males of this age tend to fight quite a bit, although among the Midget Whites, we haven't seen things escalate to the point of drawing blood very often. They are also vying for the attention of the females (there are naturally more hens than toms in the flock), and doing their best to emulate their dad by fanning out their tails and puffing out their feathery

Turkeys following us through the woods.

white chests. Their attempts at mating lack a certain finesse at this stage. I suspect that the boys are also trying to impress us, knowing that we will be keeping only a few of them for breeding next spring. Okay, maybe I'm giving them too much credit there.

Since a rash of bobcat attacks a couple of years ago, we have been reminded of how much we appreciate the relative lack of trouble the turkeys have had with predation. I was nervous the first year we raised Midget Whites, especially when they were small; I worried that when they began to free-range, their bright white color would make it too easy for predators (particularly hawks and eagles) to spot them. We think that part of their survival is due to their tendency to hang out as a group most of the time. They also don't usually wander as close to the edge of the woods as the chickens do. Say what you like about their brains, we love their instinct for survival!

Then there's the wonderful dance the toms do when trying to impress the hens, and the curious deep rumbling sound — like the sound of a Harley-Davidson starting up in the distance — you can hear it if you're close enough to a tom when he's expanding his chest. We're fascinated by the way a tom's face and wattles change color rapidly, from deep solid red to a pale pink to bluish-white. What's this all about? We laugh about it being like some kind of built-in mood ring, or maybe it's a hormone thing. And of course, the way they walk around the house, looking in all the windows trying to see where we are; they're such adorable little Peeping Toms.

As we gear up for the two days of slaughtering right before Thanksgiving, we're thinking over all we've learned and experienced over the past several years of raising turkeys. They are sweet, often funny and entertaining creatures. While they may not be the most intelligent animals around, they have definitely been an asset to our farm, and it's hard to picture life here without them.

CHAPTER 12

Chicken Tractors and Slug Slurpers

H AVING FOLLOWED US SO FAR ON OUR JOURNEY up the steep side of the poultry learning curve, perhaps you can imagine our mental wheels turning when we first encountered the magical phrase "chicken tractor." Although I did later read Andy Lee and Patricia Foreman's delightful book of the same name (see Appendix B), I don't remember now where we first heard this concept mentioned. I just know that I wanted to learn more. The sad fact that I had not yet located a fully restored mint-condition John Deere Model B tractor for less than $150 may have been a factor, too.

So, I consulted my usual sources: magazines, library books, the Internet. I had no trouble finding lots of photos, descriptions, even actual plans for building a "chicken tractor." I was thrilled, I tell you. Thrilled.

What exactly is a chicken tractor? Basically, it is a smallish coop on wheels, usually with an enclosed run extending from the front of the coop. Ideally, the width of the coop and run match the width of your garden beds, say three or four feet at most. The idea is that the birds descend from their elevated roost in the morning and spend their day happily foraging in the garden bed, confined and protected in the run. Each day, the coop can be moved along the bed, which is tilled, turned, cleared of slug eggs and weed seeds and, of course, fertilized.

Seeing as how we don't have children to delegate our chores to, we both thought this a wonderful idea. However, we've concluded that, for the most part, chicken tractors aren't especially practical in our situation; yet another thing we had to learn by trying it for ourselves. First of all, since there is space to allow our birds to free-range, we prefer to let them do that. Second, our flock size increased rapidly over the first couple of years (remember the slippery slope), and we were selling eggs before our first year raising chickens was out. My sense is that chicken tractors would be ideal for the smaller-scale poultry grower, for example those who live in urban or suburban areas. Certainly it's a good idea if you have many potential predators around.

We have built and use several mobile coops fairly similar in design to chicken tractors. I like that the birds are in a coop off the ground at night, adding another layer of security against nocturnal predators. If a bird needs to be temporarily confined when it is injured or sick, one of the smaller coops is a handy place to isolate it for a few days. The first of this type that I built — fondly known as the Deuce Coop — is traditionally used as a transition coop for chicks that have been brooded indoors. It has removable adjustable roosts and a roof that is high enough to hang a Coleman lantern from if we think the little birds need some extra warmth at night.

Deep Bedding: Pros, Cons and Our Experience

We have tried using the popular deep-bedding system in most of our coops at one time or another, with different birds and different kinds of bedding material. Theoretically, the idea is that you start with a few inches of bedding on the floor of the coop. Then you add layers of bedding at regular intervals; ideally this is timed so that, by the time the cold weather sets in, there is a good thick layer of semi-composted bedding in the coop. As the mass of bedding increases, it begins to generate heat and composting begins, killing various kinds of parasites that inhabit coops, as well as providing supplemental heat. Proponents of this system claim that, done properly, the deep-bedding system allows you to make coop cleaning a once-a-year chore. What's not to like about that?

Our experience has been that the deep-bedding system works better in larger coops than in smaller ones. In hindsight, this seems logical, since the system is dependent on a certain minimum mass of bedding in order to encourage actual composting to begin. However, before we tried it ourselves, not a single article I read on the subject mentioned anything about the size of the coop being a factor. Our largest coop, the turkey house, is six feet by six feet inside, and the largest chicken roost area is four feet by eight feet. We tried deep bedding in both of these coops, and it worked reasonably well. Still, we cleaned the coops out more than once a year, and eventually went back to the old system, which works well for us.

It is certainly possible that inadequate or poorly designed ventilation is a factor with the deep-bedding system. To me, even though we now have ten coops regularly housing poultry, it is not a major inconvenience to simply clean out the coops on a regular basis. If we can smell even a hint of ammonia in the coop, it means there is too much moisture for the bedding to absorb. The cleaning process gives us the opportunity to inspect the coops and roosts for mites or other potential problems that can occur in coops. It is not unusual to find eggs laid on the floor of the coops, which would be wasted (not to mention possibly attract rats or other pests) if the coops weren't being inspected or cleaned regularly.

Honestly, most of our coops are small enough that it only takes ten or fifteen minutes to clean them out anyway. Some are easier to clean than others — I've been on a learning curve with coop design too — but none of them are so difficult that it discourages us from cleaning them out often. During the winter especially, when the short days mean the birds are spending long hours in their coops, it's amazing how fast droppings can accumulate and start producing ammonia, which can lead quickly to eye and respiratory problems. Why take the chance on this when it is a problem so much easier to prevent than to solve?

You might be wondering what we do with all that bedding and manure we clean out of the coops. One way or another, it all gets composted. Bedding from the chicken and turkey coops usually gets dumped in a designated compost space in one of the gardens, normally on a bed that I'm not planning to grow anything in until the following season. We also put compost in an area that the birds have access to; they have a great time scratching through the composting bedding, competing for the worms and bugs and seeds, spreading it around and generally having a good time.

The bedding, being thus aerated and spread out, dries out quickly even in cool weather, so there is no noticeable odor after a day or two. Usually when we're ready to plant something in one of these compost areas, the bedding and manure are well broken down and easy to incorporate into the soil. In fact, it seems to work well as a winter mulch, since no matter how we pile it up, it isn't long before the chickens have spread it out over a large area. They also eat whatever seeds they find, including weed seeds, so over time, the garden beds become progressively more weed-free. It certainly saves me some time and effort too; raking heavy piles of bedding and manure is a tiring job.

Also, on the subject of the deep-bedding system, I would definitely not recommend trying this with ducks. Their poop is so incredibly wet that it quickly saturates bedding, which becomes quite heavy in the process. The resulting high moisture content of duck coops is not easily offset by even the most excellent ventilation. The best advice for maintaining duck coops is to allow plenty of floor space for the number of ducks that you have, use an absorbent type of bedding such as wood shavings and clean those coops out frequently. Once a week is good.

I mentioned before that David wanted ducks to help with slug control. I heard a great quote about slugs: "There is no such thing as a slug overpopulation. What you have is a duck underpopulation." Ducks certainly are enthusiastic slug slurpers. I am, however, apt to cringe when I see them greedily snatch up a slug, tilt their heads back and let those slimy little things slide right down their throats. How they eat things like that and produce such delicious eggs is a complete mystery to me. But they clearly love them, so who am I to be squeamish? Besides, I love not having any slugs in my garden!

Ducks Really Do Love Slugs

About a year ago, David and I had been discussing the possibility of increasing our laying duck flock (then only ten layers and two drakes), realizing that duck eggs were becoming more and more popular around here. So after we got a call from a local farmer who wanted to get rid of some organically raised laying ducks, we debated only briefly before calling him back and accepting his generous offer. Although he had said there were a number of drakes in the flock of about thirty-five, we figured we would be at least doubling the number of our laying ducks. Also, the thought of bypassing the five-month process of brooding and raising baby ducks before starting to collect eggs had an obvious appeal for us.

So we folded down the back seats in our Subaru wagon, lined the whole back area with a heavy tarp and a thick layer of straw and headed down the hill to catch us some ducks.

We were so thankful for the dry weather that December day, although it was cold and very windy there, because the ducks were in a yard that was all down to mud. They move quite quickly, and these smallish Khaki Campbell and Indian Runner ducks were good fliers as well. We moved the fence netting around to create a small corner, on the theory that we would drive a few ducks at a time into the corner, close it off with a section of fencing and grab them.

It worked reasonably well. One thing about ducks, they like to do everything as a group. And these particular ones were very nervous, understandable since we were obviously strangers to them. Ducks also truly hate to be handled, so we were pleasantly surprised at how calm they were once we had actually picked them up.

It took nearly an hour and a half to catch them all, but we got them one at a time and put them through a back-seat window into the car. And although it turned out that there were forty ducks in all, they seemed to have plenty of room back there.

When we finally got on the road for the half-hour drive back home, the ducks were understandably a little bit anxious, and they were all quacking at once, as they will do. All things considered, though, they were fairly calm during the drive. We knew that ducks like to be talked to (and even sung to), and we wanted them to get used to the sound of our voices, so we kept talking. At one point, David said, "Hey, anybody want to stop for ice cream?" A few ducks quacked. David tried again. ☛

"How about tacos?" This suggestion also generated some unenthusiastic quacking. To my offer of fried chicken, there was no response at all. Finally, David called out, "Who wants some *slugs*?" Whereupon a loud chorus of excited quacking erupted from the back of the car. We just about fell off our seats laughing.

I told you ducks were smart, didn't I?

Ducks in Subaru.

CHAPTER 13

Hunt and Peck:
Putting Natural Foraging Behavior to Work

ONCE WE GOT OUR FIRST DUCKLINGS, shortly after our first turkeys arrived, we added up the score: Over a hundred chickens, twenty-one turkeys and now nine ducks. As you can imagine, all those birds were going through a fair amount of feed. I'm sure being on pasture and having lots of bugs and other tasty things to eat was making a dent in the feed requirements, although this seemed truer of the older birds than the younger ones. The youngest, those in the brooders and the Deuce Coop (the "halfway house" where they learned to roost before making the transition to "big kid" coops), had their own higher-protein starter feed for their first six weeks. Then they switched to poultry grower mash, and the young turkeys had their own grower formula. It was a lot to keep track of, especially since our feed was delivered about once a month from a company in British Columbia, so we had to order ahead of time.

Early in our third year of poultry farming, it occurred to me to try growing some grains. I wasn't thinking that this might make a huge difference in the amount of grain we had to buy for feed. Heck, I didn't even know if it was going to be possible to grow things like oats or sorghum at our elevation of a thousand feet. Also, since most available pasture was already being grazed by a whole lot of free-ranging birds — and deer — I didn't have much space available for experimenting.

Six-week-old Midget White turkey.

Once again, I started reading. And planning. I was surprised to find that much of the really practical information about growing grains was in seed catalogs, such as those from Bountiful Gardens and Territorial Seed Company (see Appendix B). And once again, eventually I got to the point of putting the books down and giving it a whirl.

It was quite an experience. Not only were we fairly successful the first year, on that oh-so-familiar learning curve, but it was a whole lot of fun taking down the fences (put up temporarily to protect the crops as they grew) and watching the birds excitedly harvesting their own lunch. The turkeys are tall enough that they could easily reach the grain heads on the oats, but the chickens had to jump. Presumably they were blissfully unaware of how funny they looked doing this. I'm not sure if they ever figured out that, if they just let the turkeys have first crack at it, enough of the grain would fall to the ground that they could come in after the turkeys and simply clean up the fallen grain. It was entertaining, though.

As time goes on, I certainly hope that we can gradually begin to grow more of our own feed grains. The drought in the Midwest in 2012, which resulted in a shortage of corn and correspondingly higher feed prices, has brought the subject back to the table for us. We have limited space to devote to growing grain, and we continue facing the challenges of a short growing season and our elevation. But in order to begin to meet our long-term goals for self-sufficiency, we feel we simply have to put more effort into it.

Soaking Grain Mash for Poultry

Just this past year, I had an idea, one that, in retrospect, I wondered how in the world I hadn't thought of it before. We have been feeding our birds various kinds of organic grain mashes for years. (Note: formulated pellet and crumble feed is commonly referred to as 'mash' or 'grain'. By 'grain mash' I mean a mixture of ground pure grains and legumes.) Often this feed has a certain amount of very finely ground particles and lightweight chaff. The problem was that the birds would eat all the larger pieces and turn up their collective noses at the more powdery leftovers. Also, no matter what kind of feeders we used, the birds always seemed to fling a fair amount out onto the ground. In the winter, especially, when the ground is often wet, this feed would quickly go mushy, the birds wouldn't clean it up, and it would get to be a smelly mess.

The solution came to me one day when I was mixing up the mash for our pigs. We always soaked their mash and corn in water for several hours before feeding them; they did a better job of cleaning it all up, and I suspect they may also have found it easier to digest that way. (David endeared himself to the pigs by soaking their morning mash in hot water; "hot cereal," he called it. "Fine," I grumbled, "Just as long as the things aren't getting spoiled or anything.")

Anyway, it finally occurred to me to soak the grain for the birds the same way. It took a bit of experimenting to find the right amount of water to add to the grain — too much, and the mash was runny and the birds didn't like it; too little water, and it was hard to mix up and didn't solve anything. Once we figured out the proportions, though, it was great. The birds ate up all their feed; even the tiny particles soaked up the water. The feed wasn't getting kicked out on the ground. With feed prices as they are, I was pleased to no longer see so much feed getting wasted. ☞

Another thing I noticed after we started soaking the feed was that the size of the eggs was increasing a bit. We keep production records, so it's easy to spot trends. We started collecting more extra-large eggs and fewer large ones. I wonder if soaking the grain is somehow making more of the nutrients available to the birds. I do know that the extra effort of soaking the grain before feeding it out has had excellent results.

(**Note:** I wouldn't recommend you try this with pelleted or crumble-type feed. I suspect you'll end up with a mushy mess. However, if you do try it and your birds love it, do let me know.)

Chapter 14

Turkeys and Chickens as Guard Animals?

JUST WHEN WE THOUGHT WE HAD OUR TURKEYS FIGURED OUT, we had yet another new experience with our Midget Whites early one fall morning. Around 9:30, I was in the living room, about to indulge in a cup of Earl Grey tea and my uncontrollable daily writing habit. David had gone upstairs to take a short nap. It is usually fairly quiet at this time of day.

After just a few minutes, the turkeys started making noise in the front yard. Most of the flock of nineteen was out there, and they were making that particular noise that we've come to recognize as their warning that a predator was on the ground nearby. (Usually when we've heard this sound and checked it out, it turned out to be deer. The turkeys make a very different noise when an aerial predator is overhead.) I stood up, looked out the window and could see the flock on high alert, standing still with necks all stretched upward in the same direction, loudly sounding the alarm. I figured there were deer on the other side of the split-rail fence, munching on vetch; lately we've seen a doe and her two tiny fawns frequently in that area.

I opened the front door to get a better look. Since David was trying to sleep, I was hoping to get the turkeys to quiet down. I couldn't believe what I saw: a good-sized black bear, just at the north end of the front yard, about sixty feet from where I stood. It was walking right up into the yard, and all the turkeys were just standing there, squawking loudly, making no move to get away.

After calling up to David, I grabbed my camera and headed quickly upstairs to a better vantage point for photographing the bear before it left the yard. The bear, hearing me yelling to David, turned away and headed back toward the road and the edge of the canyon. I was just in time to get one decent photo before it disappeared into the trees by the edge of the canyon. By then the turkeys were beginning to calm down.

I, on the other hand, was having a bit of an adrenaline rush. This was only the third time I have actually seen a bear up here; the other times one had been down by the pond, a couple of hundred yards from the house. Our property is bordered almost entirely by State land, with our nearest neighbor two miles down the hill, so naturally

Rooster Talk: Predator Alert or Photo Op?

The other day, David and I ran outside when we heard an aerial-predator alarm from our roosters. I don't know how to describe this exactly, but we've started to discern some subtle differences in this kind of alert. This time, our first reaction was that it must be a big bird coming in low overhead. Sure enough, we got outside just in time to see a pair of golden eagles cruising low overhead, the first one wasn't much more than 50 feet up, certainly close enough to shake up the birds.

They were enormous, amazing, beautiful birds, and it was the first time we've seen golden eagles here, although lots of bald eagles have been around in the past year. Silly me, I had run outside without my video camera. David said to me later that, for birders like us, the roosters' alarm calls can be helpful in letting us know not just that there's a potential predator in the vicinity, but there might be a rare opportunity to see something like a golden eagle. They usually are higher up in the mountains than our farm's elevation of a thousand feet.

As I said, there seem to be some variations in the aerial-predator call. I'm going to try to record these someday. In the meantime, I must remember to grab my camera next time I hear our helpful little roosters give the heads-up.

we have lots of wildlife around. It was very surprising to see a bear so close to the house, especially in the middle of the morning.

It seems to me that it wasn't until recently that we became aware of the turkeys' excellent early-warning systems. We were familiar with the roosters calling out to the hens when an aerial predator was about, warning them to get under cover (an amazing sight to witness), but the turkeys are very consistent in letting us know when a visitor arrives on foot. We've even seen the turkeys chasing deer and wild ducks out of the yard! How they distinguish between our ducks and the wild ones I don't know, but apparently they do.

As we head deeper into the autumn, and the food sources so plentiful in summer become less abundant, we will be keeping our eyes open for hungry predators who show interest in our birds. And we're very grateful for the efforts of the turkeys to help in that process.

CHAPTER 15

Weasels Are Smaller Than You Think

ABOUT A YEAR AFTER WE STARTED RAISING CHICKENS, we experienced our first major predator attack. Eight little "teenagers," chickens about nine weeks old, were killed in their coop one night. Since we were closing them in their coops at night, we had assumed they would be safe. After the initial shock and disbelief eased a bit, we tried to figure out what had happened, as we had no idea what had killed the birds, or how it had gotten into the coop. Everything looked all right from the outside, but inside the coop, all eight of the birds were obviously dead, and it was a bloody mess.

We concluded that it had to have been a weasel. At the time, we had twenty-one young turkeys, as well as several broods of chicks, and apparently this had attracted the attention of some of the local bad boys. I remember being amazed to learn that a weasel can get through a space as small as one square inch. Obviously, weasels are smaller than I thought. The only weasel we've seen around here is the tiny Least Weasel; picture a jumbo hot dog with a tail, and you'll have a pretty good idea of how small this creature is.

At that point, I was beating myself up pretty hard for not knowing this earlier. It's all very well, I thought, to learn from my mistakes — I had built the coop in question — but it was horrible to learn that lesson at the expense of those innocent little lives. It was very difficult to keep from imagining what it must have been like for them; in the dark, not knowing what was happening, only that a destructive force was in the coop, rampaging its way around the roosts until every last bird was dead. Ugh.

That same day, I spent about two hours going over and over that coop, making sure that there was no space left that even approached one inch wide. However, the first night after that when different birds were in that coop overnight was nerve-racking for me. I made up my mind to work hard at learning all I could about designing and building coops that would meet the birds' need for safety, health and comfort.

I'm still learning.

I've noticed in poultry yards that I've visited, as well as in magazine photos, how many coops seem to be more or less slapped together out of whatever scrap materials are on hand. I don't say that critically, but merely to point out the real problem: It's hard to find, especially locally, a decent coop for sale. Even if you do find one, it's expensive, and not very likely to be the right size for the number of chickens you need to house. Many of the coops I've seen at local feed stores are only large enough for three or maybe four adult chickens.

So for us, and probably for many of you as well, the best option seems to be to build your own coop. There are plenty of plans out there, and even entire books on the subject (see Appendix B). But what if those plans are for coops that are bigger or smaller than you need? Maybe you only have a few chickens now, but you think you might have more at some point in the future. What are some general guidelines for good coop design?

In my experience, two things are essential for ensuring the health and well-being of your poultry, no matter how many you have. One, keep those feeders and drinkers clean. Two, make sure you give them plenty of room.

The first of these guidelines is self-explanatory. The second covers more ground. By plenty of room, I mean floor space inside the coop; room to move around, flap their wings, dust-bathe and scratch and peck outdoors without stepping on another bird in the process; plenty of feeder space and adequate roost space and for the number and size of your birds.

If you have ever raised day-old chicks, you probably know that, by the time they reach about four or five weeks old, they start looking around for something to roost on. But there's more to providing roosts than simply nailing a dowel in place. Here are some guidelines.

First, how much roost space do you need? This depends on how many birds you have and which breeds they are. Larger birds need bigger and stronger roosts, as well as enough space for them to sit comfortably side by side. A common mistake is not making the roosts big enough. The diameter must be large enough so that your chickens' feet don't curl all the way around. When chickens roost, they settle down on their feet, covering their toes with their feathers to keep them warm. If their toes go all the way around and under the roost, they won't be completely covered by that nice down blanket. This can lead to frostbitten toes in cold weather. (More about this in Chapter 31.)

There is no rule that says roosts need to be round. We have had good results using two-by-two and two-by-four lumber for roosts. Two-by-twos (actually 1½-by-1½ inches) work well for most chickens. We use two-by-fours (1½-by-3½ inches) for turkeys, with the narrow side up when the birds are young. Turkeys, even smaller breeds like the Midget White, have large feet and long toes, and they need strong support when roosting. When they get older, we turn the two-by-fours so the wide side is facing up.

And what about bedding? For the first few years, we used straw, which is readily available and generally cheap. Combined with poultry manure, it makes a valuable addition to a compost system. We recently ran up against an issue with straw, though, when we applied for organic certification. It seems the rules (in Washington State) say that straw used for bedding must be certified organic. The problem was I couldn't find any organic straw for sale anywhere near our farm. After spending quite a bit of time on the phone and the Internet, I found the closest source was nearly two hours away, including the ferry ride. For the amount of straw that we used, and considering our limited storage space, I couldn't figure out a way to make it cost-effective to go this far to buy it.

So, reluctantly, I began to consider alternatives. At this point, the best option we've found that meets the organic standards is wood shavings, sold as pine or "white" shavings. As long as they are from untreated wood, this complies with organic requirements; it's relatively inexpensive and easy to work with.

Since late summer of 2012, we have been using white shavings for bedding in our coops. It's a little too soon to know for sure how it compares as far as composting goes. I would much rather be able to use straw in the nest boxes, as the chickens seem to find it easier to make nests in straw; they also tend to kick the shavings out of the nest boxes. I'm hoping that they will get used to the shavings, though. I do believe that the shavings are more absorbent than straw, a major point in their favor for me. They also seem to be a little easier to rake out of the coops than straw; straw tends to mat down and turn into a soggy, heavy clump, and the shavings appear to stay a little looser. I must admit I like using the shavings for bedding more than I had expected.

After much experimentation with coop design, I've come up with an idea that has worked really well for us. It consists of three parts: A simple coop, with adjustable roosts, a side door for easy cleaning and the option to add a nest box; a lightweight run that can be pushed right up against the front of the coop; and a "trolley," a wheeled run sized to hold the coop and the lightweight run. We've found this to be a really simple, flexible system capable of accommodating chickens of all ages.

Here's how it works. When we brood chicks indoors, we need a transitional coop to house them for a couple of weeks when they are moved outside. We put the coop on the ground, set up the beginners' roosts, hang a Coleman lantern if the nights are cold, and it makes a

Four-week-old pullets.

dandy brooder. After a few days, or sooner depending on the weather and how well-feathered the chicks are, we push the light run up to the coop and open the door to give the chicks access to fresh air and grass. We move the coop and run every day so they are always on clean grass. You might be surprised how quickly even little birds can graze down the grass and make their yard a big poopy mess.

Once the chicks are six weeks old or so, we lift the coop and run up onto the trolley. At this point, the coop system looks like a chicken tractor. It's easy to move; the coop is up off the ground and secure from predators. Both ends of the trolley are removable, so when the birds are big enough to free-range, we take them off during the day and the birds can come and go as they please. The coop is designed so that, by simply removing two boards at the back, a nest box can be quickly added when the pullets reach laying age.

Some of our coops are stationary; most of them are mobile. I've gotten better at designing and building coops over the past few years, but there are lessons yet to be learned. I've never forgotten what I learned as a result of that weasel attack. We know that, short of keeping the birds constantly confined, we won't always be able to prevent losses. But we're working hard to keep them as safe as we can.

Ramps or Platforms?

The first year we had chickens, we lost two laying hens. I discovered the first, lying dead under the coop overhang, one day when collecting eggs. Of course, I had not the slightest idea what had killed the poor bird, but it was obvious something had. Except for her head and neck, which were bloody and torn up, I couldn't see any other sign of injury. We talked about it a bit, and buried the hen.

The next day, once again out collecting eggs, I lifted the nest box lid and surprised a very small skunk, caught in the act of dispatching another laying hen, right on her nest. Unfortunately the bird was already dead, and the nest box was a bloody mess. Fortunately I didn't get sprayed by the skunk.

This was upsetting, and not just for the obvious reason. I admit, to that point, I had been somewhat complacent regarding predators; this was the first time we had lost any chickens to them. I had assumed that, other than hawks and eagles, all potential predators were nocturnal hunters. Since we closed up the birds in their coops at night, I took it for granted that they were safe.

Additionally, it was clear that the skunk had simply walked right up the handy ramp into the coop and into the nest box behind the roost area. The type of skunk we have here, the Spotted Skunk, is much smaller than the common striped skunk. It never occurred to us that something like a skunk would just walk right into the coop like that.

We solved the problem by taking away the ramp and replacing it with a platform attached to the coop in front of the door. The birds only have to fly (more like a jump, actually) up sixteen inches to the platform, but this height is too much for the little skunks to reach.

This was one of those lessons that we learned the hard way about keeping our birds safe. Fortunately it was a quick and easy fix. Still, it was our first loss to predators here, and a tough lesson it was.

CHAPTER 16

Bobcats with Chicken Breath,
and Other Bedtime Stories

A COUPLE OF YEARS AGO, we experienced several exciting days, in a man-versus-jungle sort of way. After seeing a large bobcat out near the edge of the canyon in the morning, we later saw what appeared to be a mother bobcat and one of its babies. Evidently it was the day to train young Bobby to hunt. David actually saw the youngster grab a chicken and try to run off with it. The cat was so small that it had to hold its head up high to keep the chicken from dragging on the ground, hampering its escape attempt by affecting its ability to see where it was going. David made a loud noise close to the kitty. Young Bobby immediately dropped the chicken (which was apparently unhurt other than losing a large wad of feathers) and ran into the nearby berry bushes.

Meanwhile, I was stationed, with my camera, about forty yards away on the south side of the old black walnut tree, expecting the mother bobcat (with or without young Bobby) to head that way. Sure enough, Mama came out of the brush about twenty-five feet from me, just the other side of the walnut tree. She saw me right away, changed directions without missing a beat and bolted off across the shooting range into the bushes on the north side of the hill. Alas, I wasn't able to get a photo; she was simply too quick. I live in hope, however.

Looking around the area where young Bobby had grabbed the pullet, we discovered several piles of feathers, all looking like they had come from our young New Hampshire pullets. We decided to do a head count that night after the birds were all tucked in, to try to determine how many we might have lost; we were hoping it was very few. We did realize, though, that we needed to do some serious strategizing as far as predator control was concerned.

One useful thing we did was to start educating ourselves about the hunting habits of the local wildlife. Bobcats, we learned, like to hunt at the edge of the woods, sneaking up on their quarry and staying under cover until the last minute, then jumping out to grab the unsuspecting prey. We've actually witnessed this, even in our front yard; the birds start squawking, and we take a look, usually just in time to see a cat jump over the fence, snatch a chicken in its mouth then leap back over the fence. It's amazingly quick, and honestly, we have to admire the beauty and grace of these animals, even if we don't always appreciate their choice of lunch entrées.

Keeping in mind their hunting habits, we began a major effort to clear away the brush, low-hanging tree branches and all the nettles, bracken ferns and other vegetation that grows like crazy during the mild, wet spring weather. We figured we could make it somewhat harder for the cats to sneak up on the free-ranging birds by taking away their cover. Since we really don't want to completely confine the chickens and turkeys in fenced areas, this seems to be our best strategy. It's helped greatly just to do some reading on the subject, to understand the hunting habits of bobcats and other predators; we now feel like we're not completely at the mercy of the cats and hawks. We also plan to hatch more birds in the spring, to make up for occasionally sharing some with the native wildlife.

One Saturday morning, there were three bobcat attacks on chickens (that we know of). I witnessed one, which happened so fast I didn't have time to get a photo. Good grief, are these cats speedy! Ironically, at the time, I was starting to clear out some blackberry and thimbleberry bushes, in an area where we suspected the bobcats were hiding out, waiting for some clueless foraging chicken to stray too near. Hearing the turkeys start up their ground-predator call, I looked up just in time to see a large bobcat jump out of the bushes and tall grass about 60 feet from me and grab a New Hampshire pullet. I have to admit that, although I was naturally upset that another of our birds was attacked, I also felt in awe of the beauty of this animal and the speed and efficiency of its attack and retreat.

Around the same time, David was down in Sequim, taking another of our hens to the vet. It had clearly also been attacked by a bobcat and had a deep laceration across its shoulders. David had tried to close the wound with Superglue (seriously), but it was too difficult to get the moist edges of skin to stay together. It was a valuable laying hen, so he took it in to get stitched up. The vet also gave us a solution to use to clean the wound several times a day.

I ran into the house and called David right after seeing the attack, and he headed home. I went back out, with something of an adrenaline rush going on and my camera in hand, on high alert for more signals from the turkeys that the cat was still around. I assumed that while I was inside telephoning, it probably had run back into the woods, although I was gone barely two minutes. When David got back and had put the stitched-up hen inside, he came back out, and we decided to try to flush the cat out of hiding, if it remained in the bushes.

It took only a minute. I was on the south end of the area in question, David on the north. He started hacking his way through the berry bushes with the Swedish brush hook, and suddenly a large bobcat emerged from the bushes about thirty feet from me. It saw me and pivoted, racing away to the west toward the edge of the canyon, the most likely place for its den. Although my camera was ready to shoot, I only managed to get a very blurry photo; man, that thing was fast! I also noticed that its feet made no sound at all as it ran. Amazing creature.

Shortly after this, I noticed one of our two New Hampshire roosters walking a bit awkwardly. He also seemed to have some loose feathers around his shoulder. Being somewhat hypersensitive to bobcat attacks at the time, I had David catch the rooster so we could examine him. Sure enough, puncture wounds on both his shoulders! The poor thing was clearly in shock. We brought him inside, and for the first day or so, we didn't know if he would make it.

At this point, we looked at each other and decided that, whatever else was on the agenda that day, we needed to make it a priority to do what we could to protect our birds. Over the next two days, we cleared an amazing amount of berry bushes and low-hanging tree branches, mainly around the edge of the canyon and the bush-covered area where I had seen the latest attack.

We don't expect to completely eliminate the problem of predation, of course, but we will continue to do what we can to make it harder for the cats to sneak up on the birds.

Goshawks Prefer Organic Chicken, Too

Last fall I was able to get some up-close and personal photos of a Northern Goshawk at our farm. Unfortunately, it was standing on the dead body of one of our young New Hampshire cockerels, but what can you do.

We speculated that the bird was a young adult male goshawk; it appeared to be almost completely into its adult plumage. And like most raptors, the male is smaller than the female, and this one couldn't get off the ground with the dead chicken. (The adult goshawk tops out at 2.1 pounds, and this fourteen-week-old cockerel was easily twice that weight.) As I approached, the hawk tried to move the chicken, but it only managed to drag it a few inches across the grass. I wonder if the hawk learned a lesson from this.

I had been trying to get photos of a goshawk for a while now. Our friend Shelly Ament, a wildlife biologist with Washington's Fish and Wildlife Department, told us that goshawks are not often seen by humans in the wild. (Ha, I thought. Try letting some chickens free-range in *your* backyard. Bet you'll see some goshawks.) I retrieved a primary wing feather, which I saw fall while the hawk was flapping its wings, and we saved it to give to Shelly. She had asked us to be on the lookout for goshawk feathers. Her department is interested in comparing the genetics of goshawks in our area to those of hawks in British Columbia, so she was pleased to have the feather.

We understand that since we choose to free-range our chickens, turkeys and ducks, we're likely to lose birds to predators from time to time. However, we feel fortunate to live in a place where we sometimes see beautiful animals like the Northern Goshawk. What a gorgeous bird.

Above: *One of our tiny, gorgeous Nankin roosters.*

Below: *Four-week-old Ameraucana chick.*

Above: *Cochin banty with Midget White poult.*

Center: *New Hampshire and Ameraucana eggs.*

Below: *Heritage chicken parts in stockpot .*

Above:
Ducks splashing in the morning sun.

Center: *Three-week-old Khaki Campbell ducklings.*

Below: *Snow doesn't bother these hardy ducks.*

Above: *Blue Swedish
duck splashing in
roasting pan.*

Below: *Khaki Campbell
drake.*

Opposite: *Blue Swedish
duck.*

Above:
*Mom turkey with
babies on fence.*

Below:
Turkey preening.

Opposite:
*Young turkeys
show off their
flying skills.*

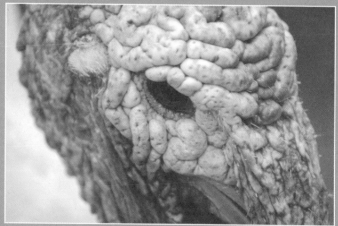

Above:
Young Midget White turkeys on the fence.

Below:
Closeup of Old Tom.

CHAPTER 17

Keeping the Ducks Safe at Night

WE HAVE TWO PONDS HERE AT THE FARM, and one is a large natural peat bog. It is down the hill, below the woodshed to the east of the main house and bordered on three sides by trees. Its situation makes it a natural sanctuary for many kinds of wild birds; to date, we have identified sixty-three bird species, many of them migratory.

Of course, being a good-sized body of water with plenty of natural cover nearby, it is a magnet for ducks, including our own. Because we let the ducks free-range during the day along with the chickens and turkeys, we expected that, sooner or later, they would wander down the hill and find the bog. The question was what would they do then? Would they come back up the hill to get their dinner and be tucked in at night? Or would they decide that life on the water was pretty nice for ducks and camp out?

Some of both, it turns out. There were times when we would just stand at the top of the hill by the woodshed and call them (I'm not kidding), and they would come marching up the hill within a few minutes. Then there were the nights when David would spend a lot of time in the rowboat, chasing them around the pond. This frequently led to the use of colorful language and threats, usually involving firearms. It was nice to tell ourselves that they would eventually come back when they got hungry, but we also knew that they could last for quite a while on the rich frog population in the bog.

Besides worrying about the ducks being safe when they camped out, we also were obviously not able to collect eggs from them. This was especially disturbing when we had young ducks just starting to lay; once they get in the habit of laying somewhere away from the coops, it's almost impossible to get them to change their routine.

Then, one day, an amazing thing happened. For some unexplained reason, the ducks, who had been happily racing down the hill to the bog every morning as soon as the coop doors are opened, started coming up the hill, on their own, about half an hour before they would normally be hunkering down for the night. We no longer had to stand at the top of the hill, calling and singing to them (I'm not kidding about that, either). The ducks were just coming up the hill, gobbling down some grain, drinking water as if they haven't been on a pond all day, and letting us escort them to their coops.

What a relief! We love our ducks, some of which we've had for over four years now, and we just want to keep them healthy, happy and safe. We're very grateful now, as we tuck them in and tell them a little story, that they've figured out that we just want what's best for them.

CHAPTER 18

Turkeys in the Canyon

J UST ANOTHER QUIET MORNING ON THE FARM. I drove down into
Sequim to deliver eggs to the Alder Wood Bistro and got back up
the hill a little after 9:00. As I was approaching the house (it's about
half a mile in from our gate), I noticed one of our turkeys standing by
the roadside at the edge of the canyon. I didn't think much of it until
I got into the yard and parked the car. That's when I realized that
something was wrong. Usually when we, or anyone else, drive up, the
whole flock of turkeys comes running up to greet us, surrounding the
car and making loads of noise. Did I say it was a quiet morning? It
was a bit too quiet!

Part of our property is a canyon. As you drive in from the gate, on
the right it drops off three hundred feet to Canyon Creek. Much it is
very steep, for instance in the area where I saw that turkey. As we had
lost a number of birds (mostly chickens) earlier this year to bobcats
and cougars, it really made me nervous to think of nineteen turkeys
down in the canyon, where the wild cats hang out during the day.
You'd think someone had put a neon sign out there advertising an all-
you-can-eat turkey buffet, and our turkeys had invited themselves.

And of course this had to happen on a day when David was in Seattle.

I went inside, dropped the mail on the kitchen table, put the tea-
kettle on to boil and went back out to look for the turkeys. I headed
over to the spot where the turkey was when I drove in, but it was no

longer in sight. I walked to the edge of the canyon, looked down, and sure enough, there was the whole flock of silly birds, roosting in the low-hanging branches of cedar trees. Did I say silly? Suicidal is more like it! Feeling silly myself, I called to them, informing them of the certain doom that awaited them if they didn't get back up into the yard post-haste. Some of the toms gobbled enthusiastically at me, the hens simply ignored me, but none of them showed any interest in being obedient. Fine, I thought. I'll just go have my cup of tea.

Then, about fifteen minutes later, the turkeys all showed up at once in the front yard, with not a bit of remorse among the lot of them for all the anxiety they had caused. The good thing about turkeys is they seem to do everything as a group; once one of them decided to leave the canyon, they all did.

So what if my tea is cold? I'm just glad I didn't have to hike down into the canyon to retrieve a bunch of wayward turkeys. Especially ungrateful ones.

Turkeys love roosting in trees.

CHAPTER 19

Equal Rights for Unhatched Chicks

I F THERE'S ONE THING I HATE, it's taking eggs away from a broody hen. You know what I mean. You open the nest box lid, and she fluffs herself up indignantly, hisses warningly at you and generally makes no secret of her deep resentment when you reach under her to collect the eggs she has every intention of hatching. Then there's that pathetic little peeping sound she makes, usually just before you find the last egg she has defiantly hidden under her wing. "I just want to hatch my babies!" she protests. "It is my prime directive! You're a woman, can't you understand that? You wouldn't possibly be so cruel as to deny me the fulfillment of my heritage henhood! Would you?"

Oh man, I hate that. Nothing like a little guilt trip every time you pick up the eggs. And yet, this strong tendency to broodiness is one of the reasons we love our heritage birds. They are not only willing, they are positively eager to incubate and hatch eggs for us. This is great, of course, at a certain time of year — breeding season comes to mind — but when a hen goes broody in earnest, she stops laying. Most of the time, we want to keep the egg production up, so we try to discourage the hens from being broody by taking away their eggs.

Easier said than done. Theoretically, if you collect eggs frequently, and don't allow a hen to accumulate a clutch, she won't really go broody. This is another thing about poultry that sounds good on paper, but in real life, it doesn't always work. It depends on the hen;

some are more easily talked out of it than others. It depends on the time of year, although some hens are inclined to go broody any time more than one egg is within a hundred yards of them for more than a few minutes. It can even depend on how much privacy she feels she has in her chosen nest space. For all I know, it has something to do with hormones or the phase of the moon. We've had hens who were so determined to be broody that we simply had to take their eggs away several times a day and wait for them to get over it. Sometimes even desperate measures, such as furtively whispered threats involving sharp knives and stockpots, have not the slightest effect.

During the spring breeding season, naturally we like it if a few hens go broody, as long as the timing is right for collecting hatching eggs. Managing hens who are incubating eggs, though, does involve a fair amount of time and organization. For one thing, each hen needs her own private space, especially if you have many laying hens. This usually means having her own coop, or at least a nest box that can be more or less isolated from other birds. She needs her own food and water, access to sunlight, grass and a place to dust-bathe. Broody hens don't like to potty in their nests, thank goodness, so they need to be able to hop off their nest and go somewhere else once or more a day to poop. She needs to be protected, not only from predators, but from the other birds, who are insatiably curious and will, given the opportunity, always investigate a nest with a clutch of eggs. You definitely want to avoid this, as it frequently results in broken eggs in the nest.

Here's an example. A couple of years ago, one of our ducks set up housekeeping in the salal bushes not far from our woodshed. By the time we had realized she was missing and found her nest, we guessed she had been out there for more than a week; she had a full clutch of eggs. Boy, did she hiss when we came anywhere near her nest. The weather was warm, and the nest wasn't far from the house, so we decided to just let her be. We did make sure she had ample food and water nearby, though.

The duck, like all broody birds, was presumably hopping off her nest once a day, at least for a few minutes, to poop and grab a quick bite to eat and some water. Well, apparently during one or two of these temporary absences, a chicken managed to sneak in there and deposit one of her eggs on the duck's nest. You may remember that chicken eggs are incubated for twenty-one days, duck eggs for twenty-eight. At a time when we were guessing there was a week or so before the ducklings were due to arrive, David came up the hill with news. "That broody duck just hatched two chicks!" he said, with a strange look on his face.

"You mean she hatched her ducklings, right?" I responded, mentally doing the math and wondering if we were simply mistaken about how long the duck had been out there incubating her clutch of eggs.

"No, I mean she hatched two *chicks*," he repeated slowly and a little more loudly, as if he were attempting to explain to someone of some other native language. I must have looked confounded, as he sighed theatrically and motioned for me to follow him down the hill.

I'm sorry to report that I had a hard time not laughing hysterically when we arrived at the poor duck's nest. She was looking a little bewildered, standing next to the nest that was still full of unhatched duck eggs. In the middle of the clutch, scrambling awkwardly over the large duck eggs, were two chicks, one light yellow, the other almost completely black. The poor mama duck, who for all I know is capable of doing mental math herself, must have wondered what the heck was going on.

The good news was that one of our hens was fairly broody, and we managed, surprisingly easily, to convince her to adopt the two chicks and brood them. We couldn't resist calling the chicks Donald and Daisy.

It can be a problem, though, if a hen gets into another hen's nest once incubation has begun. Unless you mark the eggs in some way — we sometimes write a date in pencil on the eggs in a hen's clutch — if another hen lays eggs among the other hen's clutch, they will naturally not all hatch at the same time. We've found it's easiest to simply isolate broody hens once they have a full clutch and seem prepared to go the distance toward impending motherhood.

Almost every year that we've had poultry, we've had one or more unexpected arrival of a new family. This is just one of those things that we've come to accept as occasionally inevitable, since we choose to let our birds free-range during the day. The tiny Nankin banties love to lay their eggs away from their perfectly nice nest boxes; it's not unusual for a Nankin hen to come marching out of the bushes with a batch of tiny chicks a few times a year. Last year, one showed up early one chilly fall morning with ten teensy chicks huddled under her. Another time, I discovered, quite by accident, a Nankin hen incubating a clutch of eggs under the edge of a tarp covering a compost pile in my garden. The Midget White turkeys also love to make nests in the shrubbery and tall grass away from their coops. We have had more than one family of ducklings hatched *au naturel* as well.

In one sense, we don't mind when this happens. From a purely selfish point of view, it saves us the minor hassle of finding housing for everyone when more than one hen is broody at the same time. For their own safety, when we do find a hen on one of these "wild" nests,

Nankin hen with brand-new chicks.

we certainly prefer to move them into a broody coop for the rest of the incubation period, but this is not always possible. Some hens take exception to being disturbed in this situation, and may actually refuse to continue the incubation after they are moved.

As you might expect, most of the poultry mating action happens in the spring and summer months. I mentioned the Midget White turkey hen who hatched her babies on New Year's Day. Fortunately we found her nest and were able to move her into a broody coop near the house. She could come and go as she pleased, but she and her eggs were out of the winter weather, not to mention safer than they might have been in her nest at the edge of the woods. Another time, a hen showed up early one morning in late November with sixteen hungry chicks in tow. We certainly were scrambling that day to find adequate broody coop space for the exhausted hen and her large family.

It might seem as if we're not keeping track of things very well here, but as I said, this is part of the trade-off of allowing the birds to free-range. As I write this, we have close to a hundred and fifty birds here, housed at night among ten separate coops. It is impractical to do a head count every single night, so if a hen camps out once in a while, we don't always know it unless we just happen to see her in the trees or bushes. This nesting behavior is instinctive for these birds, and while we obviously don't actively encourage them to nest in the bushes or the woods, we aren't all that inclined to discourage it either. We might feel differently if it happened more often, but so far it hasn't been a major problem.

Over the past several years, we have bought three batches of chicks, New Hampshires and a few Ameraucanas. We had decided to ultimately maintain a dual-purpose laying flock of these two breeds (everyone seems to love the blue and green Ameraucana eggs). It seemed clear to us that we would be better off buying chicks, rather than hatching our own, while we were in the process of transitioning our mixed flock to these two breeds.

There were several reasons for this. First, we had established a poultry business in our community. If we simply got rid of all the birds in our flock of breeds that we didn't ultimately want, the flock would dwindle quickly to a point where we would not be able to continue supplying eggs to our customers. Second, to successfully breed and hatch purebred chicks in a mixed flock, we would need to isolate breeding groups for several weeks at a time. This would mean building even more coops and fairly intensive management. Third, we hoped that by buying several batches of chicks, each batch from a different hatchery, we would be able to achieve a greater degree of genetic diversity within our flock. Three seasons later, we now have, other than the Nankin banties, only three hens that are not either New Hampshire or Ameraucana. These three are over four years old now, so we will no doubt either sell them this spring or slaughter them as stewing hens.

Of course, from this point on, it makes sense from a sustainability perspective to breed our chickens and supply our own replacement

Nankin chick.

birds. Frankly, after brooding three batches of chicks indoors for several weeks, I'm more than ready to let the hens take over the job. We've found our New Hampshire hens, while not as broody as the Nankins and Cochins, are easily persuaded to incubate their eggs.

Speaking of brooding, we have learned a few things about this process. Most significant, to me anyway, is the length of time chicks need to be brooded, that is, provided with supplemental heat while they grow out their feathers. Conventional wisdom always seems to advise the same thing: Start out the brooder at 95°F for the first week, then drop the temperature by 5°F every week until the brooder is down to room temperature. This, to me, is another example of a "rule" that is generally accepted without question, even though it doesn't take into account factors such as time of year, type of bird and so on. You would think that if you lived in the mountains and bought your chicks in March, you might need to brood them for a longer period than if the same chicks were hatched in June.

Our actual experience has been that even day-old New Hampshire and Ameraucana chicks seem comfortable at temperatures around 75°F, as long as they are protected from drafts. We also believe that when brooded at these relatively cooler temperatures, the young birds feather out more quickly. As a result, we rarely have needed to brood our chicks indoors for more than about two weeks before they are moved to their transitional outdoor brooder. This has been consistently true for us no matter what time of year it is.

Oh, I am looking forward to breeding season this year. Even though managing broody hens does take a fair amount of effort and extra coop space, these hens are wonderful mothers. Just having someone else taking over the brooding chores makes it worth the extra work to me. At our farm, when a hen hatches chicks, she usually has them off the nest and out walking around on the grass with her within a couple of days, pretty much as soon as all her eggs have hatched.

I admit I am always a bit nervous about these teeny tiny chicks. The turkeys and other chickens are attracted to their cute and incessant peeping noises, and it would be so easy for a much larger hen or turkey to step on a chick. Not that I have actually ever seen this happen. The mothers are quite protective, and they seem to know when their babies are cold or hungry or have simply had enough, and they head back to the nest together. One of these days, I'll probably figure out that the hens know what they're doing.

Make It Easy to Keep Your Brooder Clean

Here's a useful idea that we came up with that has worked well for us when brooding chicks indoors. Put a layer of bedding in the brooder, white shavings or clean dry straw, and lay old pillowcases or sheets over the bedding, completely covering the brooder floor. This has several advantages. One, if you sprinkle some starter feed and baby grit (don't forget the grit!) on the pillowcases, the chicks readily learn to peck and eat. Even day-old chicks show a strong instinct to scratch and peck, a delightful thing to see.

Two, covering the bedding prevents the chicks from eating and filling up on wood shavings or other non-food particles. Turkeys are particularly bad about doing this. Young birds can actually die from lack of nutrition if they habitually eat bedding, which they may do even when other food is easily available.

Three, it makes the brooder a whole lot easier to clean out. I lay the pillowcases so that one overlaps the next; it's quick and easy to simply roll the whole thing up, starting from one end, and replace it with clean cloths. Washing out the poopy pillowcases in a bucket of warm water always makes me think of my mother washing out cloth diapers when I was young. This is so much easier than trying to change out the actual bedding in a box of forty or fifty chicks, trust me. And at the rate that this many chicks eat and poop, believe me, you want to make it as easy as possible to clean out that brooder.

CHAPTER 20

Keeping Poultry with Other Farmyard (and Backyard) Animals

S EVERAL YEARS AGO, we got our first pigs, two seven-week-old pure-bred Tamworth gilts. The Tamworth, I had read, is considered to be the rooting and grazing champion of the pig world. They apparently like nothing better than to dig up every blade of vegetation in sight, chewing up the plants, roots and all, with great enthusiasm and appetite. Having also read that the root word (no pun intended) of "pig" is the same as that of "plow," I was intrigued.

Hmmm, I thought, eying a large flat meadow full of six-foot-tall canary reed grass, I wonder what would happen if we turned the pigs out on this stuff and let them graze it down and plow it up? Then we could plant some clover or timothy or something and turn this into a really nice pasture.

Well, to make a long story short, the Tamworths did justice to their reputation. As they got bigger, they were plowing things up so rapidly and efficiently that we had to rotate them onto new pasture every few days. What we didn't realize at the time was that our poultry would also play an important role in this pasture improvement project.

We noticed that, as we moved the pigs from one grassy area to the next, the chickens and ducks immediately moved in to investigate the freshly upturned soil. We subsequently learned that not only do the birds break up and spread around the manure piles, encouraging speedy composting, in the process they also eat parasite eggs and weed seeds that pass through the pigs' systems. This helps break up the life cycles of various kinds of worms and other parasites that tend to affect pastured birds and other animals. We were amazed.

In a truly sustainable system, all types of poultry can peacefully — and quite usefully — co-exist with other animals, whether that is the family dog or a herd of milking cattle. If you already have one or more animals or pets and are considering starting to raise poultry, you might want to look over the Poultry from Scratch worksheet (Appendix A). Answering the questions will help you decide if poultry might work well for you and your other animals. In the meantime, here are some of our thoughts about poultry and peaceful co-habitation.

Dogs and Wildlife

You may have heard that domestic dogs kill more chickens than any other predator. We know a few people who have dogs and chickens, and just about all of them swear that their chickens grew up with their dog, the dog loves them, and the chickens aren't the least bit afraid of the dog. I think that's great, and I hope it continues to work out for them. I am not a dog person myself — I don't dislike them so much as I'm just not used to being around them — but I suspect that when problems do occur between dogs and chickens, it is when two or more dogs get together and the dynamic changes. Most people I know who live on farms or have poultry have one or more dogs in residence. In fact, people often wonder why we don't have dogs ourselves, especially considering all the birds we have free-ranging here during the day.

There are two main reasons why we don't have dogs. First, neither one of us wants to have a dog. Second, we live in a place that is surrounded by many, many acres of woods. This area is rich in native wildlife, and our view is that we are living in their backyard — not the other way around — and it's up to us to figure out how to make it work for us and our birds. We believe that a dog would scare away the wild animals that live here. I know that sounds paradoxical, since most people who have dogs for protection have them just for that reason: to scare off the wildlife. I can only say that I think the wild animals were here before we were, and they have every right to not be driven from their territory just because we want to let our poultry out on pasture during the day. Yes, we've lost some birds to predators, sometimes as a result of ignorance on our part, and more often because we can't be everywhere all the time watching for hungry wild animals. We also feel that we ought to do all we can to preserve the environment of our property, which happens to be wildlife habitat. If we ever get to the point where we feel that predation losses are so high that the birds are stressed out all the time, then we may just decide that it doesn't work to have poultry up here.

This seems like a good place to say again that, in my view, time spent in preparation before bringing home chickens or turkeys or ducks is fairly essential. This includes learning what predator issues you face in your area, and whether your situation might require you to keep your birds confined in order to keep them safe. It is the responsible thing to do.

Cats

We have one cat at the farm: Sir Winsome de Cosmos, or Cosmo for short. He is fortunately quite happy to be an indoor cat, most of the time. Occasionally if a door isn't latched, Cosmo will hook a paw around the edge of the door, pull it open and head outside. More recently, though, he only seems to do this after dark. I suspect this is because the last time I remember him sneaking outside during the day, the turkeys spotted him from across the yard and immediately all raced toward Cosmo, chirruping loudly with their ground-predator alert call. Poor Cosmo took one look at the rapidly approaching turkeys and sprinted right back into the house. I tried not to roll on the floor laughing in front of him; I swear I tried. Ever since this incident, he has not seemed the least bit tempted to go outside during the day, even when the door is wide open.

We've had Cosmo for two years now. The first year, when he was going outside in the daytime more often, he never showed the least inclination to go after our birds, not even the tiny Nankins.

He tended to hang around behind the woodshed, a place where our birds don't go very often; apparently he was only interested in stalking small wild birds. I'm not sure if he ever actually caught one, but he seemed to enjoy the process anyway. And as long as he didn't pose any threat to our poultry, the only concern I had with him being outside was that he would be grabbed by a predator himself. So I'm happy to have him be an indoor cat. Even when he does slip outside at night, he usually stays near the door and is content to come back in after a few minutes, especially if it happens to be raining or snowing.

On the other hand, Cosmo is a terrific mouser. We are in and out of the house frequently, and there are plenty of opportunities for small rodents to find their way inside. This is especially a problem during cold weather. Cosmo has made it his personal mission to keep the house rodent-free, which is obviously a blessing.

Other Livestock

Besides the birds, pigs are the only other livestock we've raised ourselves, so I can't speak from experience about how cows or horses or sheep or goats interact with poultry. However, I do believe that, when you have animals in a backyard or small farm setting, diversity is nearly always a good thing. I've heard more than once of people successfully grazing chickens along with cows. I've also heard stories about intruder-chasing guinea hens and geese that graze on garden weeds while supposedly ignoring your vegetable plants. Do your own reading. Participate in online forums. Talk to your friends and neighbors and the guy at your local feed store. I've found farm people consistently generous with their time and expertise when it comes to lessons learned and practical advice.

One More Thing ...

Another important issue to consider when you have other animals and want to raise poultry (or vice versa) is time. It's easy to overlook this factor, but believe me, it makes a difference. You might think it won't take much time each day to care for a few chickens or ducks. Probably, most of the time, you'll be right. But who is going to clean out the coops? If you have dairy animals, will the milking schedule conflict with poultry chores? If you have a job away from home, how are the eggs going to be collected regularly? This doesn't need to be a big deal. However, I would encourage you to spend a little time and effort figuring these out first, discussing the issues with your family, and I'm confident that both you and your animals will benefit in the long run.

CHAPTER 21

Small Farm New Math:
If (Chicken Tractor), Then (Pig Plow)

A COUPLE OF YEARS AGO, on a sunny early-December morning, I drove to Sedro Woolley, a small town in beautiful farm country, nestled snugly at the foot of the Cascade Mountains north of Seattle. I visited a lovely organic farm, on a mission to buy some Tamworth pigs to add to the two we've been raising this year. Normally I wouldn't have been buying pigs at that time of year, but these were being offered at such a good price that I made the trip without a second thought. The eleven-week-old pigs were bigger than I expected. It was a challenge getting the three of them into the two dog carriers I had brought, but finally they were tucked in and settled for the three-hour drive. Once we were on the road, the three little pigs slept pretty much all the way home.

The previous year, we had decided to diversify our small farm by adding two pigs. Although I had read extensively on the subject, I felt just about as ignorant as I had prior to starting with poultry. I was also excited, though; it was something new and different.

I had heard that the Tamworth breed was particularly noted for its rooting ability. Never having been in close proximity to pigs, other than at a county fair and the grocery-store meat section, I had only a hazy idea of what "rooting" actually meant in real life. Well, as far as the Tamworth is concerned, it means they will plow up pretty much everything within reach of their sharp little hooves and long strong snouts. Our first two Tamworth weaners were seven weeks old when we got them; we were amazed to see how quickly and efficiently such little pigs turned a grassy pasture into loose soil.

We haven't had our rototiller out of the shed since.

You might be wondering what pigs have to do with poultry. Well, as I said, our first motivation in getting pigs was to diversify our farm operations. I also love to cure prosciutto, pancetta, bacon and other pork products, and in our area, there are limited choices in commercially available pork. We figured, chicken tractors are great, so why not pig plows?

David and I decided that, since these pigs wanted to root all day long, by golly, we'd put them to work doing what they love. The sizable area to the east of our main house, between the shooting range and the peat bog, is six or seven feet deep in reed canary grass in the summer. It is also the largest plot on our property that could potentially be turned into good pasture. With a high water table, it is essentially self-watering. It's flat, gets good sun in summer and, unlike most of our 40 acres, has no trees. As we watched the piglets happily tossing large clumps of sod in the air, the wheels started turning. What if we could transform this previously unused acre into prime grazing land?

We had just moved our two older pigs off this area, as it was fairly wet down there after all the rain and snow around that time. They had done their job beautifully, though, rooting and leaving behind an expanse of thoroughly tilled, peat-rich soil; all it needs is a bit of leveling, and it will be ready to plant. With the high water table, I will probably opt to plant ladino clover and possibly timothy, both crops that can deal with having wet feet some of the time. We are looking forward to seeing that field transformed into lush pasture over the next season or two.

In the meantime, the pigs are happily plowing up their new yard. The three little pigs (I know, I know) are in a separate yard temporarily, while we train them on the electric fence. They also are enthusiastically rooting and grazing. They all look happy and healthy, and appear to be enjoying their typical routine: Eating, rooting, grazing and napping.

Such is the cycle of Tamworth life. We did the math, and we like the results.

CHAPTER 22

Farm-fresh Eggs: What's the Big Deal?

I DON'T KNOW ABOUT YOU, but before we started selling eggs in 2008, I was frankly baffled by the terminology of egg carton labeling. Although some labels are periodically redefined, others remain consistent, for example, egg sizes. What's so confusing about large and extra large? Well, nothing, once you know what it means.

Egg sizes are determined by weight. The standard large egg is 2 ounces. However, the difference between one egg size and the next is ¼ ounce; thus, a large egg is actually anywhere from 2 to 2¼ ounces, an extra large egg is between 2¼ and 2½ ounces and so on. Sometimes, because of their varying shapes, one egg might appear to be larger when they are both the same weight. So when sorting our eggs, we weigh each one to determine the size.

By the way, you cooks and bakers out there probably already know that most recipes calling for eggs specify the large size. Now that you know a large egg is 2 ounces, you'll know what to do when you want to use a different size: Weigh the eggs until you have the equivalent weight of the eggs called for in the recipe.

OK, on to **egg grades**. Just a few years ago, I had not the slightest idea what this meant. Simply, an egg's grade is an indication of freshness, Grade AA being the freshest. How is this measured? It is determined by the size of the air space at the wide end of the egg; this is easiest to see when candling (shining a bright light through the wide end). This air space gets bigger because the shell is porous, and over time the contents gradually evaporate. So, the smaller the air space, the fresher the egg.

With commercial eggs, the catch is that this standard of freshness is applied when the eggs leave the place of production, not when they arrive at your grocery store. You can't tell just by looking at the eggs, right?

Here's another egg carton label: **Free range**. This designation is unfortunately not as clear, in terms of definition, as it might appear. Technically, for eggs to be labeled "free range," the hens must be allowed "access to the outdoors." What does *that* mean? (David suggested that the hens are watching the Nature Channel while pumping out eggs.) In reality, it means very little. A commercial operation can comply with this requirement by simply installing one or two very small doors somewhere on an outside wall of their 20,000-hen facility, making sure the door is open part of the day. Honestly, now, how many of those hens, in a place that huge, ever have a chance of even seeing the door? And if they do go outside, who's to say there is grass out there? It's far more likely the door opens out onto the edge of a parking lot.

The main problem with the term "free range" is that you can't know for sure. These eggs *might* have come from birds who spent part of their day outdoors, but you simply have no way of knowing. It seems to me that without tighter definitions and oversight, "free range" isn't much more than an effective marketing tool. You'd expect to pay more for "free range" eggs, right? But what are you paying for?

Once again, it comes down to this: If you don't raise your own chickens, the best way to know that you're getting truly fresh eggs, from birds that have been treated humanely, is to either buy them at a farmers market, or better yet, visit the farm where the eggs are produced. Observe the operation. Talk to the farmer. You'll not only return home with eggs you can believe in and enjoy, but also establish a relationship that benefits you, the farmer and your whole community.

Among egg carton labels, one of my personal favorites (in a dubious sort of way) is **vegetarian diet**.

Should You Wash Your Eggs or Not?

I would not have believed how much strong feeling this particular subject stirs up. Some people assert that eggs should always be washed, preferably sanitized too. Others insist that any washing or cleaning is somehow detrimental to the quality of the egg.

If I tell you what I think, will you promise not to send nasty e-mails or (horror of horrors) un-friend me on Facebook? Okay, here goes. First let me say that, since we were first licensed to sell eggs wholesale, we have been obliged to follow the guidelines of the Washington State Department of Agriculture's (WSDA) Food Safety Program. Having also done a lot of research on this subject, I think that the WSDA's guidelines are very sensible.

Because eggs are perishable and, under certain circumstances, subject to bacterial infection, the idea is to collect, clean, dry and refrigerate them as quickly as possible. We use warm water and an old soft toothbrush to clean them. The water should be warmer than the eggs. Why? Because the shell of an egg is porous. The theory is, if there is mud or chicken poop or whatever on the egg shell, and you wash the egg in cold water, the contents of the egg will shrink away from the shell, bringing with it anything lingering on the outside of the shell. Since some eggs have just been laid when they are picked up, and a hen's body temperature is 103°F, we try for a wash-water temperature of around 110°F.

"But," you never-wash-an-egg advocates are shouting, "washing the egg removes the bloom!" I know, I know. And I am going to take my social-networking life in my hands and ask you, "So what?"

The "bloom," as I understand it, is some kind of coating that is applied to the outside of the egg's shell right before it exits the hen's body. I have occasionally picked up an egg that has been so freshly laid that it is wet; presumably this is the "bloom." Some argue that removing it results in a shorter shelf life for the egg. First of all, unless that coating is somehow completely sealing the entire eggshell, I don't see how this can be true. Remember that the shell is porous; probably if it was coated thickly with wax or something, its contents wouldn't evaporate. I don't know what the makeup of the "bloom" is, but I doubt it is actually sealing the egg to that extent. ☛

And anyway, we deliver our eggs several times a week to our customers. We *know* they are being consumed when they are quite fresh. So, frankly, shelf life is of no real concern to us.

Even if we weren't subject to the WSDA requirements, we would always clean our eggs. Occasionally an egg looks so clean that I don't bother washing it. However, I rarely find an egg to be so pristine that it can't be improved by a light cleaning. I can see no advantage in leaving mud, chicken poop or bedding stuck to an egg for fear of compromising the "bloom." And since we are talking food safety here, honestly now, why take chances?

What does this mean? According to the Humane Society of the United States, "vegetarian diet," in the egg industry, means that "these [laying hens'] feed does not contain animal byproducts, but this label does not have significant relevance to the animals' living conditions."

Okay, then! Let's set aside for the moment the reference to living conditions (which is vague in the extreme). What strikes me about "vegetarian diet" is that the marketers are counting on the average consumer to be unaware that chickens are actually omnivorous. We see our birds foraging all day long, eating all manner of worms, bugs and grubs; they go fairly nuts during the summer when, for about a week, the carpenter ants are flying. It's quite entertaining to watch the birds run around, leaping into the air to snatch the large and apparently delectable ants.

We've even seen the chickens catch small frogs, lizards, mice and the occasional snake. The first time I saw a hen with a snake I wouldn't have believed it. From across the yard, she was racing around, with ten or more in hot pursuit. The lead hen had something hanging from her beak, but I couldn't tell what it was, so I went to see what was going on. About twelve inches of the garter snake's body was visible. With an impressive finishing kick, the hen got far enough ahead of the other birds and stopped abruptly. Tilting her head back in a whiplash motion, she swallowed the snake in one go. I swear I'm not making this up. I remember wondering if the snake was dead when the hen swallowed it.

Anyway, the point is, if a laying hen is eating a "vegetarian diet," by definition, she is not foraging any of her food. True, I think it's great if the hens' feed doesn't contain any animal by-products; I just happen to also believe that hens are happier, healthier and lay better-tasting and more nutritious eggs when they can access the naturally balanced range of foods they thrive on.

CHAPTER 23

Duck Eggs Are Delicious, and Great for Baking

THE OTHER NIGHT, David and I had dinner at the fabulous Alder Wood Bistro in Sequim. As you know, the Bistro buys the vast majority of our organic chicken and duck eggs. Recently, as our young ducks have started to lay, we've had a few more duck eggs than usual. Just in time for Tapas Tuesday, Bistro chef Gabriel Schuenemann came up with a deceptively simple, outrageously delicious dish to showcase them.

At first glance, you might have thought it was a plate of scrambled eggs and toast. But wait ... this *is* the Bistro, after all! Embedded in the smooth, creamy egg were shavings of Washington State black truffles, and the first bite confirmed our initial impression: The eggs were positively swimming in butter. Nestled alongside them were thin crispy crouton slices large enough to pile a couple bites of truffly eggs on top. The combination of super-fresh, melt-in-your-mouth tender eggs with crunchy croutons and quantities of warm butter sliding seductively down your chin ... oh, boy!

Duck eggs aren't always easy to find, but if you do, it's well worth trying them. We're often asked what the difference is between duck and chicken eggs. Aside from the size — our duck eggs average over three ounces, compared to the standard two-ounce large chicken egg — our general impression is that duck eggs have a somewhat milder taste and are richer and creamier. The whites of duck eggs also have higher viscosity than chicken eggs; this makes them a great choice for baking, especially in recipes where the eggs are separated. The baker at the Bistro likes to use our duck eggs for the Chocolate Bliss, a wonderful flourless, gluten-free brownie.

Over the past year, we increased the size of our duck-laying flock, mainly to keep up with the needs of the Bistro. There is also increasing interest in duck eggs around here; we've heard from a number of people who want to buy them as soon as we have any extra to sell. Although they naturally cost more than chicken eggs, no one seems to mind; their size and quality, along with relative scarcity, adds up to good food value. And like all our birds, the ducks are fed organic grains and free-range on pasture during the day.

A while back we lost several of our ducks. They usually headed down the hill to the bog in the morning, spending most of the day there, then coming back up for a bedtime snack before being tucked into their coops for the night. Around the time of a recent rash of bobcat attacks, the ducks suddenly stopped going down to the bog. We're not sure why this happened, although David's theory is that because one of the ducks was killed by some predator down there, the others are avoiding that area. Whatever the reason, I'm just happy that they're hanging around closer to the house now; besides knowing that they are safer, I find it's much easier to collect their eggs when they lay them in their coops!

When laying ducks are up to full production, they are very prolific. Khaki Campbell ducks, so we hear, can average 340 eggs per year; that's more than most chickens lay, even in their prime. And when we have more duck eggs than the Bistro can use, other customers are always waiting in the wings, so to speak.

Chef Gabriel also does amazing things with our chicken eggs. Recently he came up with a gorgeous salad that has smoked salmon, pickled onions and a poached egg on top. Delicious. Most of our chicken and duck eggs are used in their desserts, though: my favorite crème brûlée, their seasonal organic carrot cake, an incredible hazelnut torte and various seasonal fruit tarts, among others. (See Appendix B for more information about the Alder Wood Bistro.)

We love our ducks, and we love the Alder Wood Bistro! They buy so many of our eggs that our standard joke is that we have to go to the Bistro to have some of our own eggs. It's worth it, though; we never know what Gabriel is going to come up with next. Committed to sourcing ingredients locally, he's always willing to try things we suggest, such as the duck eggs. Even though we're a small farm, we're proud to have had a role in the success of our hard-working friends at the Alder Wood Bistro.

CHAPTER 24

Slaughtering and Processing Poultry

WITHOUT DOUBT, one of the more unpleasant realities of raising poultry is slaughtering. Even if you keep birds mainly for egg production, as we do, sooner or later this issue will come up, I promise. Because we breed our own replacement laying hens, we inevitably have more roosters than needed for the next breeding season. (Supposedly each batch of chickens hatched average out to half pullets and half cockerels.) Even if you buy replacement hens every year or two, you eventually have to decide what to do with the older hens. And because our birds free-range, we occasionally have to deal with one being injured in an accident or by a predator, or even in a pecking-order disagreement. Poultry raised on pasture also sometimes fall victim to disease or a poisonous plant.

Whatever the case, the question of slaughtering will come up at some point. Obviously, if you plan to raise some birds specifically to process as table fowl, you need to come to grips with the reality of what that means.

Which brings me to a confession: So far, I have not been able to bring myself to participate in the actual killing of a chicken or turkey (we have yet to slaughter any ducks on our farm). Yes, I do feel guilty about that. Poor David always has to do the hard part, while I deal with the unpleasant but comparatively easier tasks of scalding, plucking and packaging. I also am in charge of cooking up and canning any poultry that we don't plan to eat fresh, such as older chickens.

Part of my reluctance to killing birds is, I am sure, due to my general squeamishness about throat things. Anything that even conjures up images like choking or the cutting of throats instantly makes me queasy. We slaughter our birds by cutting the large veins on either side of their necks with a very sharp knife. When this technique is well-executed, a chicken will hardly seem to feel a thing, and it bleeds out in just a couple of minutes. Still, I simply have not been able to get past my instinctive reaction, although I haven't actually ever tried it. If David asked me to, I would certainly make an effort, but I have yet to volunteer.

When David slaughters a chicken or turkey, he sits on a chair, holding the bird in his lap, talking to it until it is calm and quiet. He

Slaughter vs. Harvest

We have often been asked why we continue to use the term "slaughter" in favor of the current trend toward using others such as "harvest" to describe the killing of farm animals raised for food. It's a somewhat tricky question, as we don't wish to offend anyone or seem to be passing judgment on their choice of terminology. I'll just say that there are definite reasons we choose to say "slaughter"; please hear me out and know that I respect your choice, whatever it may be.

First, the *Merriam-Webster Dictionary* definitions. *Harvest*: (1) The act or process of gathering in a crop; (2) to gather, catch, hunt or kill (as salmon, oysters, or deer) for human use, sport, or population control. *Slaughter*: To kill (animals) for food.

Do you see the distinction? "Harvest," by the second definition, is clearly referring to wild animals, those "caught" or "hunted" as opposed to those specifically raised for food. Also, we feel that using "harvest" in the context of killing chickens or turkeys for food (vaguely grouping this process with "gathering in a crop") is frankly euphemistic.

Our position, then, is simply that we feel more comfortable using the term "slaughter."

then makes a swift, bold cut with a razor-sharp knife on either side of its neck. The whole process is carried out with as much consideration for the bird as possible. If that sounds odd, given that the bird's life is about to end, consider this: If the chicken or turkey is shaking or seems fearful (this happens more often with turkeys), David usually lets it go and we try again another day. We both feel tremendous respect for all of our poultry, and we try hard to show that at every level of our husbandry practices. From our point of view, these beautiful animals are our partners, active participants in a process that continually sustains us and our environment. For many reasons, they deserve our best efforts to ensure they have a happy, healthy, natural existence and a painless and dignified death.

You may be wondering why we don't simply chop off their heads. Many people believe that is the quickest and most pain-free way. This may be possible, but honestly, I don't think it's the best choice of killing methods. First of all, if you've ever tried to hold a live chicken or turkey in position on a chopping block or tree stump, you'll know not only is it awkward for you, but the bird will likely struggle and be stressed and fearful. This just makes the whole process even more unpleasant for all involved.

In addition, when the head is cut off a live bird, the windpipe is obviously severed. This can result in the aspiration of blood into the body as it flaps its wings and moves even after it is dead. The aspirated blood can contaminate the meat if it isn't drained promptly.

What it comes down to for us is that holding them upright in a normal posture and cutting their neck veins is the best way to slaughter birds in a respectful and relatively stress-free way.

The first time we slaughtered birds was a few months after we got our first chickens. Although I had done some reading on the subject, we both had very little idea of what it would be like. We were not surprised that the reality involved details that weren't mentioned in any of the articles I had read. At the end of that first slaughter day, I couldn't help thinking that it would have been helpful to have known some of these points.

For example, nothing I read had indicated that we should be prepared for the chicken to flap its wings energetically — after it was already dead. We were slaughtering three roosters that day, and it was incredibly stressful because we realized, right in the middle of slaughtering the first, that we had no idea how to tell if it was actually dead. It was horrible. The rooster kicked and flapped and jerked when we thought it must already be dead. We were afraid that it was suffering, and we didn't know what to do to fix it. The only way we knew for a fact that the birds were dead was when they had bled out completely and we had removed their heads prior to scalding and plucking.

Another detail I remember about that first slaughter day was the smell of wet feathers. While I was dunking a bird's carcass slowly up and down in the hot water of the scalding pot, I found it impossible to get away from that smell. It was nauseating. Since that day, I've gotten more or less used to it, however unpleasant, especially with turkeys.

I want to pass on one aspect we've learned about slaughtering, as I think it's important: Find a quiet place to do this job, if at all possible out of the range of vision of other poultry. We have found that chickens and turkeys especially are instantly attracted to loose feathers and to blood. If they are allowed even momentary access to the slaughter area once the killing has begun, they will be in the way, pestering you and getting underfoot all day long.

When we have turkeys to sell before the holidays, we always let our customers know when they will be slaughtered so they know when to pick them up. We have no way of freezing large birds like turkeys, so we generally slaughter on the Monday and Tuesday before Thanksgiving, and customers pick them up on Tuesday or Wednesday. What has been surprising to us is that the customers often want to come up and help on slaughtering days. One year nearly all of our paying customers showed up to participate. Obviously, having extra hands to help is wonderful. Even Midget Whites are fairly large birds, and plucking especially can be time-consuming. And of course, the turkey slaughter happens in late November, when it's usually good and cold outside up here in the mountains. I think it's great that people seem to want to learn how to slaughter their own birds, and David is skilled and an excellent teacher. I feel guilty that I have always passed the buck and expected him to do the nasty part. Maybe this year will be different.

CHAPTER 25

The Best Laid Schemes ....

But Mousie, thou art no thy lane,
In proving foresight may be vain:
The best laid schemes o' mice an' men
Gang aft a-gley ....

> — Robert Burns, "To a Mouse, on Turning Her up in
> Her Nest with the Plough, November, 1785"

Normally our pre-holiday turkey slaughtering is planned for the Monday and Tuesday before Thanksgiving. Back in the fall of 2010, I spent extra time on planning and organizing this thrilling event. We were expecting several people to come up and help with the slaughtering, some of whom were going to process the turkeys they were buying from us. We were also listening with some apprehension to the weather forecasts: The barometer was falling, as was the temperature. The chance of snow falling, however, was rising. While the predictions were for "snow showers," with possible accumulations of between one and three inches, we both felt it would be prudent to have a plan B in case conditions worsened either before or during the slaughtering.

The Sunday before Thanksgiving, light snow fell most of the day, ending in early evening, with maybe an inch on the ground. It cleared up that night, with temperatures dropping to the mid-20s,

Profile of Old Tom on a sunny day.

and remained so when I got up Monday morning. Great, I thought, no more snow overnight. People won't have trouble getting up the hill. (Snow showers, I reminded myself, one to three inches.)

Well, about 7:30 that morning, it started snowing again. A little before 9:30, our friend Giles drove up, bringing a large bronze turkey he had raised and planned to slaughter here. At that point, it was snowing harder; about three inches were already on the ground. We were pretty busy then with the final preparations: I had a big pot of homemade minestrone on the stove and a kettle of water keeping hot for coffee and tea. The wood stoves were stoked, and extra firewood was in the rack. Outside, the large screen-house tent was set up with the plucker and a table inside, and gloves, paper towels, knives and sharpeners, and cutting boards in place. The twenty-gallon scalding pot was heating up on the propane burner outside the tent, and a large bucket of ice water stood ready for cooling the freshly scalded birds prior to plucking.

It was our third year raising and slaughtering turkeys, and the first time we had had snow to deal with at slaughter time. Being late November, it's usually cold, and of course the days are short, which is

Turkey tracks in the snow.

why we usually spread the process over two days. This time we hoped to get all twelve turkeys slaughtered on one day, since we were expecting helpers. I had also done most of the setup on Sunday, thinking it would help us to get an earlier start on Monday.

Shortly after Giles arrived, we started getting phone calls. It was snowing just as hard in Sequim and Port Angeles, and in the end, we weren't surprised that Giles was the only one who made it to the farm that day. By late morning, it was getting windy, snowing steadily, and the visibility was very poor. Because of these conditions, the guys decided to get three turkeys ready for eviscerating at once, instead of doing them one at a time. (We usually do the killing, scalding and plucking outdoors, then bring the birds inside to do the eviscerating.) With the snow now piling up fast, I had to keep knocking it off the tent about every ten minutes. The footing was very slippery, which slowed us down, too. Finally we took the three turkey carcasses inside, grateful for a chance to warm up (the high temperature that day was 22°F), get a bite to eat and get out of the snow for a while.

While David and Giles worked on cleaning the turkeys, I gulped down a cup of tea before heading back outside to check on the birds' feeders and drinkers. It was so windy that it was snowing sideways, and the birds' feeders and drinkers were accumulating snow, even though they were under shelters. I also had to top off the drinkers with warm water, as the water freezes fairly quickly when it's this cold.

Since the snow had piled up to nearly twelve inches, I decided to go knock the snow off the tent again. I looked over toward the slaughter area, and guess what: The tent had disappeared. Between the wind and the weight of the snow, it had just come down. I went back in to report this to David, and we all agreed it was a good thing we hadn't been in the tent at the time! Although the snow continued to come down thick and fast, we were able to raise the tent again pretty easily. However, given the wind, the time and the cold, we moved the plucking operation indoors.

By then, it was nearly 3:00. Knowing that there would only be about another hour of daylight, we decided not to process any more birds. That evening, I phoned most of our turkey customers to let them know their turkeys might not be processed in time for Thanksgiving. Also, given the amount of snow on the ground, and the swiftly dropping temperature, it seemed questionable whether anyone would be able to get here to pick up their birds. Everyone was very understanding and looking forward to getting a fresh turkey sometime in December.

That Monday night, our low temperature was 7°F. (As David said, "You know it's cold when you have to bring the coolers with the turkeys indoors to keep them from freezing overnight.") On Tuesday, David and I slaughtered two more turkeys. Thankfully, it was sunny most of the day, although the temperature got no higher than 22°F. We knew it would get pretty cold that night if it stayed clear, and it did; our low temperature was 0°F. We processed one more turkey on Wednesday morning, loaded it onto a sled and walked the half mile to our gate to meet the customer, in case she couldn't get through the deep snow between there and the house. Fortunately the road crew had plowed our road up to our gate, so she was able to make it that far.

We definitely learned a lot from this experience. Although I had spent extra time planning and organizing, as the snow piled up, all our plans went pretty much out the window. With snow coming down hard that Monday morning, I had said to David, "I suspect the key to making it through the day will be flexibility." That certainly turned out to be the case!

On the positive side, the turkeys that weren't delivered in time for Thanksgiving put on a little more weight, since about another three weeks passed before they were finally slaughtered. We only sell our turkeys fresh, partly because we have no means of freezing them here, and also because we believe that fresh is preferable to frozen.

Once we had a chance to catch our breath and relax, we sat and enjoyed the gorgeous view: the frosty-looking Blue Mountain to the southwest, the beautiful snow-covered cedar and fir trees around our property and our two pigs playing happily in their pasture, seemingly oblivious to the weather. Although the days before Thanksgiving didn't go as we had planned, it all turned out all right, and we are truly thankful that we are safe and warm here at home in the beautiful Olympic Mountains.

CHAPTER 26

How to Cook Your Heritage Turkey

I F YOU'VE EVER COOKED A HERITAGE-BREED TURKEY, you probably noticed that it's a bit different from cooking a large Broad Breasted bird (think Butterball). Generally, heritage turkeys are smaller, leaner and with proportionately less breast meat. We've also noticed that they are a somewhat different shape; they are more elongated compared to the Butterball types, with longer legs, which can make it a challenge to find a roasting pan to fit the bird.

I won't go into the question of stuffing here, but I do want to talk about brining, which in my experience makes a noticeable difference in the roasting process. Here is an explanation for this, from the wonderful book *Charcuterie: The Craft of Salting, Smoking and Curing* (see Appendix B):

> Brines, more so than dry cures, are an excellent way to impart seasoning and aromatic flavors. A brine penetrates a chicken or pork loin rapidly and completely, bringing with it any flavors you might have added to the salty solution [garlic, onion, tarragon, pepper]. Chefs often use brines for pork, chicken and turkey, the three types of meat that benefit most from brining, because they result in a uniformly juicy loin or bird that's perfectly seasoned every time.

Roasting a brined chicken or turkey and hitting at just the right point of doneness is easier than with an unbrined [turkey]. You can actually overcook it, in fact, and it can be juicier than a perfectly cooked bird that wasn't brined. The brine seems to allow the breast to withstand the high temperature while the slowpoke legs and thighs continue to cook.

Important note: You'll need to plan ahead if you're going to brine your turkey. In addition to the actual brining time, allow extra time for chilling the brine before putting the turkey in the brining pot. Also, once the bird is removed from the brine, it needs to "rest" in order for the salt to equalize through the meat; allow a resting time of about the same as the brining time for best results. (Extra rest time doesn't hurt anything; it's better to have it rest longer than to shorten the rest time.)

Sound complicated? It's not, really. Here's how it usually shakes out, if you're planning for Thanksgiving: Make the brine on Monday and chill it overnight. Put the turkey in the brine on Tuesday and leave it overnight. Remove the turkey from the brine Wednesday morning, pat it dry with paper towels, then let rest until you're ready to roast it on Thursday.

Brine for Turkey

1 gallon water

1 cup kosher salt

½ cup sugar

Seasonings (I like to use celery, onion, garlic, parsley, carrot, peppercorns, tarragon, thyme and lemon zest)

Bring all ingredients to a simmer in a large pot, stirring to dissolve the salt and sugar. Remove from heat and chill.

You may need to double this recipe, depending on the size of the turkey and the container you use for brining.

If your turkey is on the smallish side, you can use a large stockpot for brining. A six-gallon food-grade bucket with lid also works well. This time of year on the farm, it is cold enough outside to keep the brining pot outside, so I usually plan to put the turkey in the brine before we go to bed. It stays good and cold and is ready to remove from the brine in the morning.

An important point on brining: The turkey needs to be kept completely submerged in the brine. This can be done a number of ways, depending on how big your brining container is. When using a stockpot, I put a plate inside the pot on top of the bird. Then I weight it down, either with a gallon-size Ziploc bag or a clean Mason jar filled with brine and tightly closed. If you have a very large turkey, you'll have to experiment to find the right amount of weight to keep it submerged.

Which brings me to the amount of time needed for brining. For a turkey of ten to fifteen pounds, allow twenty-four hours. If it's smaller than ten pounds (like our Midget White hens), twelve hours is about right. If it's more than fifteen pounds, allow up to thirty-six hours. Don't forget to leave time for your turkey to rest between brining and roasting.

Now, to the cooking part! If you have access to a smoker, I encourage you to try smoking your turkey. It takes longer, as usually the temperatures in the smoker are lower than your oven; on medium heat, my smoker generally cooks at between 200 and 225°F. It's hard to say an exact amount of time for smoking, as there are lots of variables (size of the bird, smoker temperature, outside temperature, etc.). Just allow plenty of time and check the internal temperature of the bird every so often. I strongly recommend brining the turkey if you're going to smoke it.

In spite of its recent popularity, I am not in favor of deep-frying heritage turkeys. In my view, brining and then roasting (or smoking) results in the best-tasting heritage turkey. In addition, frying oil is expensive and invariably poses an awkward disposal issue afterward.

Here is my preferred method for roasting: Preheat oven to 450°F. Place the turkey straight from the refrigerator or cooler into the oven, then immediately reduce the heat to 325°F. After the first half-hour of cooking, baste several times per hour with pan drippings or extra fat. If you choose not to brine the turkey, basting is particularly important.

If your turkey is under six pounds, allow twenty to twenty-five minutes roasting time per pound; for a six to sixteen pounds, fifteen to twenty minutes; and for over sixteen pounds, about thirteen to fifteen minutes. If using a thermometer, insert it into the thigh, taking care not to let it touch the bone; internal temperature should be 180 to 185°F.

As is usual with all roasted meats, remove the turkey from the oven and let it rest for up to thirty minutes before carving.

That's it! There are lots of ways to cook a turkey, but this is my favorite: brining and then either roasting or smoking. If you've never tasted smoked turkey, you're in for a real treat. It is one of the most delicious things I've ever eaten.

CHAPTER 27

Tastes Like Chicken: Making a Case for Heritage Chickens as Meat Birds

I REALIZE I'M RAPIDLY HEADING DOWN THE ROAD LESS TRAVELED, but I'd like to say a few words on behalf of heritage chickens, specifically, the reasons we are sold on them as meat birds. Nearly every single person I've ever talked to about this disagrees with me. Why? Well, once again, I believe it is partly due to context (see Chapter 6). What are you planning to do with your meat birds? Run a business or simply stock your freezer? Is it more important to you to raise the tastiest bird possible or the cheapest bird possible?

I hate to burst anyone's bubble, but if you think you're going to save money on chicken by raising your own, you're likely to be disappointed. We are obsessed with cheap food in this country, and the mass-produced, machine-slaughtered, previously frozen chicken is a good example of what we tend to expect from our food supply. Let me be clear: Unless you think your time has no value, and you never buy any feed, *ever*, for your chickens, you simply cannot compete with the ubiquitous 89¢-per-pound supermarket fowl. This is even more true of turkeys, which are commonly given away shortly before Thanksgiving when you spend a minimum amount on other groceries.

"But wait," you may argue, "I'm raising Cornish Cross chickens! They will reach slaughter weight in only six or seven weeks! How much can it possibly cost to feed a chicken for a few weeks?" Plenty, if you expect your fast-growing hybrids to reach those weights in the time claimed in the hatchery catalogs. In fact, if they have continuous access to feed, these voracious eaters will do justice to every bit of food they find, provided they don't actually have to forage for it.

The hatchery catalog hype also doesn't mention that these birds do best on a high-protein feed specifically formulated for "broilers." With the exception of turkey feed, broiler feed is just about the most expensive kind of feed you can buy, and believe me, you will need plenty of it in the next six or seven weeks, even for just a few "broilers." In addition, with the Cornish Cross, depending on how many you raise at a time, you're likely to lose some now and then, birds that don't even make it to the ripe old age of six weeks. Why? The birds' rapid growth rate causes enormous stress on their circulatory systems, and it is not uncommon for them to drop dead of heart failure. For the same reason, some fast-growing hybrids experience severe leg problems during their short lives.

A while back we heard of an old-timer who claimed a rule of thumb about the Cornish Cross was "Butcher 'em or bury 'em." He was referring to the high mortality rate among Cornish Cross meat birds that are allowed to go on growing well beyond their expected slaughter age, say ten or twelve weeks.

Obviously, if a chicken isn't likely to live, much less thrive, beyond ten or twelve weeks, you're not going to be breeding it. And as you know, hybrids are the offspring of genetically dissimilar parents; so even if you could mate two Cornish Cross chickens, the chicks could not be expected to be like their parents. Every time you want to raise more hybrid meat birds, you're going to have to buy more chicks. Keep in mind that when you buy from a hatchery, the male Cornish Cross chick will probably be the most expensive male chick of any breed.

Speaking of hatchery catalogs, I learned something I'd always wondered about: the origin of the "Rock Cornish Game Hen." When I was growing up, occasionally we would have these little birds for dinner. I remember thinking they were so exotic compared to the "regular" chicken, and I could have half a bird all to myself!

I couldn't have known that actually this exotic little fowl was a reject bird. The Cornish Cross male attains someone's idea of an ideal slaughter weight one or two weeks faster than the female. Naturally, in a culture focused on cheap food, time is money with chickens too. So the humble Cornish Cross female chick is, according to the catalog description, raised to about two pounds (live weight) and then slaughtered. Basic math estimates that the chick would then be around two or maybe two and a half weeks old. It's a relief to know that no one would be wasting any more feed on the slow-growing little things.

Becoming an Informed — or at Least Thoughtful — Consumer

Before raising our own chickens, I'm pretty sure I had never eaten a farm-fresh egg, much less a farm-fresh egg from organically raised, happily free-ranging hens. Frankly, I was skeptical. It's just an egg, I told myself. How different can one be from another?

Did I mention how ignorant I was back then? You won't be surprised to learn that I also hadn't ever eaten (as far as I know — and how would I know?) chicken or turkey raised on a local farm. I had not the foggiest idea what kind of chicken I was buying at the grocery store. In fact, I had no idea what breeds of chicken even existed, assuming there was more than one breed. Are those grocery-store chickens male or female? Did one have a choice when buying chicken? Honestly, does any of this matter?

No matter how many sources I explored in an effort to learn more, none were in favor of raising heritage-breed chickens as meat birds. What we found out about the faster-growing commercial hybrid chickens, not to mention the appalling conditions they typically live in, was motivation enough for us to not only raise our own meat birds, but to specifically choose dual-purpose heritage breeds.

Aside from the difference in the time needed to grow a heritage chicken to slaughter size, we discovered other reasons to appreciate these birds as table fare. One of the most significant, for me, is flavor. Simply by virtue of their longer lives, heritage chickens develop more flavor than the relatively immature Cornish Cross. I mentioned Julia Child's complaint about the disappearance of the stewing hen from grocery meat counters; she knew that these older birds have much more flavor. Sure, they need to be cooked slowly over a period of time, but oh, the stock that results! Dark golden brown, aromatic, with fantastic gelling power ... there is simply no comparison.

Another noticeable difference is the texture of the meat, particularly if the chicken has been free-ranging or on pasture some of the time. Why is this? Well, that bird has actually been getting some exercise. I have heard — again, I can't speak from experience here, since we have never raised hybrid meat chickens — that the Cornish Cross has to eat almost constantly in order to keep up with its warp-speed growth rate. It is apparently not inclined to roam very far from its food source, lest it faint from hunger. I would be interested to know what a Cornish Cross chicken would do if it was offered grain mash only twice a day, for fifteen minutes or so, and given access to healthy pasture the rest of the day. I suspect that, once it got used to the idea, it would do just fine, although it would probably not mature quite so quickly.

If you ever have the chance to compare a cooked grocery-store chicken to a heritage chicken that has been raised on pasture, I

Heritage chicken parts in stockpot.

promise you will notice a big difference in the texture of the meat. Often a customer has reported that one of our free-range chickens or turkeys tasted good but was "a little tough." Compared to the relatively soft and, to my way of thinking, mushy texture of the grocery-store specimen, it probably did seem that way. However, if you eat a piece of free-range chicken and then a piece of your favorite medium-well-done steak, I think you'll agree that the chicken is tender after all.

Commercial Turkey Production in the United States

Did you know that, even with the current popularity of heritage turkeys, over 99 percent of all commercially raised turkeys are Broad Breasted hybrids? In 2007, 7.87 billion (with a "b") pounds of turkey was produced commercially in the United States, a five percent increase from 2006. You would think, with numbers like this, that the turkey would get a little more respect. Um, no. In 2007, the average price received by turkey growers was a whopping 47.9¢ per pound, an increase of only .5¢ from 2006.

Mexico, where the domesticated turkey had its historical roots, imported over 3.1 million pounds of United States-grown turkey in 2006. We also export turkey to China, Russia, Hong Kong and Canada. I was surprised to learn that Minnesota grows more turkeys than any other state. Another fun fact: Israel consumes more turkey (somewhere around thirty-five pounds per capita in 2005) than any other country. That's considerably more than we Americans eat; in fact, the United States ranks third in worldwide turkey consumption, averaging a bit over sixteen pounds per capita.

CHAPTER 28

Pot Pies and Preservation

F OR ME, FEW THINGS EVOKE THE IDEA OF COMFORT FOOD more
easily than chicken pot pie. Even as a kid, I remember my mother buying those frozen individual pot pies (was it Swanson's brand?);
we simply called them little meat pies. There was something about
having a whole pie of your own, and having a choice of flavors, that
always appealed to me. Cutting into the crust with a fork, being careful not to burn yourself as the first cloud of delicious steam burst out,
knowing that it was too hot to eat and waiting impatiently as the fra-
grant gravy seeped out ingratiatingly around the edges of the crust ....
Oh, I'm sorry, I was just reminiscing.

Anyway, I love pot pies. Of course, they usually have potatoes
tucked inside, and I love anything with potatoes on principle. I seem
to go through occasional phases with food. Sometimes all I feel like
eating is anything wrapped up in a warm flour tortilla. Other times
it's anything spooned generously over a bed of brown basmati rice.
Even so, there's nothing quite like the combination of hot meat, vegetables and gravy sealed up enticingly inside a flaky, crispy brown
edible container. Carrots, peas, potatoes, onions, garlic, lots of dark
meat. (I've always preferred the dark meat of poultry, and I don't entirely understand all the emphasis on white meat. With both chicken
and turkey, my favorite part is always the dark, succulent thigh.) Meat
and vegetables swimming deliriously in a generous pool of thick rich

gravy, oh boy! My mother used to use a paring knife to scratch initials in the unbaked top crust of little meat pies to identify them: B for beef, T for turkey and so on.

There are lots of good recipes for pot pies around, from the Internet to your grandmother's recipe box. Here are two easy pie crust recipes and a basic formula for a chicken or turkey pot pie.

Hot-water Pastry

(*makes one 9-inch double crust*)

⅓ cup boiling water
⅔ cup shortening or lard
2 cups flour
¾ teaspoon salt

Mix together and form dough into a ball before rolling out. For gluten-free pastry, replace the flour with 2 cups of a gluten-free baking mix such as Manini's (see Appendix B) or use the following recipe.

Gluten-free Pastry

(*makes one 9-inch double crust*)

1½ cups rice flour

½ cup potato starch flour

¼ cup tapioca flour

1 teaspoon salt

1 tablespoon sugar

1 teaspoon xanthan gum

¾ cup shortening or lard

1 egg, lightly beaten

2 tablespoons vinegar

2 tablespoons cold water

Sift dry ingredients together into mixing bowl. Cut in shortening or lard.

Blend remaining ingredients together in separate bowl.

Stir egg mixture into flour mixture until blended. The mixture will seem quite moist, but this moisture is necessary when using rice flour.

Knead dough until it forms a ball. Divide it into two balls and roll out between two sheets of plastic wrap to desired size. To place in pie pan, remove top sheet of plastic wrap. Using the bottom sheet for ease of handling, invert the dough and drop it into the pan, shaping it into the curves of the pan before removing the remaining plastic wrap.

For later use, bake at 450°F for 10 to 12 minutes.

Chicken or Turkey Pot Pie

1 recipe pie crust dough (above)

1 recipe chicken or turkey gravy (below)

Cooked chicken or turkey meat, about a pound, boned and cut into cubes of a
 size you like

1 large onion, diced (not too small)

3 large carrots, peeled or not, cut into chunks

3 stalks of celery, including leaves, washed and sliced into smallish pieces

1 small bag frozen peas (or a can or pint-size Mason jar of canned peas,
 drained well)

Preheat oven to 400°F. In a large mixing bowl, toss meat and vegetables together. Pour hot gravy over the mixture and stir gently with a wooden spoon,
just enough to coat everything. Roll out half the pie dough and place in a deep
9-inch pie pan. Pile meat mixture into pie pan; the filling will shrink a bit while
cooking, so don't worry too much if it seems like a huge mound. Roll out the
second half of the dough and lay it carefully on top of the filling. Seal the top
and bottom crusts together at the edges. Cut a few slits in the top crust with a
sharp paring knife, to let steam escape. Put pie in oven; it's a good idea to put a
cookie sheet on a rack under the pie, just in case the gravy bubbles over. Check
on it after about 45 minutes. The pie is done when the edges of the crust are
nicely browned and the gravy is bubbling up through the slits in the top crust.
Let cool slightly before serving.

Basic Chicken or Turkey Gravy

1 stick (½ cup) butter (please, don't use margarine for this!)
½ cup flour (all-purpose flour, half whole wheat flour or gluten-free baking
   mix)
Pinch of nutmeg
Two pinches of dry thyme leaf
Sea salt and freshly ground pepper to taste
Small handful of chopped fresh parsley
4 cups (1 quart) chicken or turkey stock, preferably homemade (don't bother
   skimming the fat off the top; there's not that much of it, and you'll notice if it
   isn't there)

In a 2-quart saucepan, melt the butter over low heat; do not brown it. With a wooden spoon, stir in the flour, nutmeg and thyme, blending well. Stir for another minute or so, then gradually add the chicken or turkey stock, stirring constantly. A whisk works well to blend the stock with the roux. Continue stirring until the roux is completely incorporated into the stock; if you see any lumps, keep stirring. You still have it on low heat, right?

Stir mixture every few minutes, or constantly, if you're worried it will scorch. It will start to thicken in a few minutes and bubble a little. You definitely want to keep stirring it at this point. When the gravy is bubbling all over, turn off the heat; if your stove is electric, move the pan off the burner. Adjust seasoning to taste with sea salt and pepper. A little Tamari soy sauce (really, just a little) is a good addition, too.

Canning Chicken and Turkey Meat and Stock

Sit back and relax. I don't suppose you'll be surprised to know I have a lot to say on this subject.

I love canning. I first learned the basics of canning when I was in ninth grade. At our high school, Home Economics was required for girls in ninth grade; boys had Wood Shop, which I thought was unfair. Actually, at one point that year, the boys and girls switched places: boys had Home Ec, girls had Shop, for two weeks, if I remember right.

I didn't particularly enjoy Home Ec. Sewing was the only part I actually disliked, probably because I wasn't interested in it then and wasn't any good at it. A great deal of Home Ec was naturally about cooking. I found most of it frankly boring, as I had started to learn to cook when quite young. I already knew how to boil pasta. That same year, my mother had started making bread, and I was in the process of learning that useful skill from her. Mrs. Jones gave me a D for my apple pie, claiming I had not followed the recipe. It's true that I had used about twice as much cinnamon than the recipe called for, but I was taking the pie home, and presumably I was going to eat some of it, and I happen to like lots of cinnamon in apple pie. I should have gotten extra credit for boldness and creativity. Obviously Mrs. Jones had no imagination.

I never did understand why we made a Baked Alaska in that class, for heaven's sake. How many Baked Alaskas do you think I've made in all the years since ninth grade? Uh, let me think ... oh, that's right: *none.*

I did, however, learn the fundamentals of canning in Home Ec, for which I am grateful to this day. Certainly I've gained much knowledge and practical experience with food preservation since then, but that's where it began for me. All through high school, I helped my mother with many different kinds of canning chores: pears, peaches, raspberry and strawberry jam, bread-and-butter pickles, marinara sauce, dill pickles, probably a few kinds of jelly and other things I've forgotten.

A few years back, I had the opportunity, through our local county extension office, to participate in an intensive Master Food Preserver program. In spite of my years of canning experience, I tell you, I learned a lot. Not just about canning, either. The course covered freezing and dehydrating as well. Incidentally, it was where I learned the actual purpose of blanching green beans to prepare them for freezing, something I had always done but without knowing why. There is an enzyme in vegetables that, left to its own devices, sets in motion the decomposition process right after harvesting. Blanching in boiling water for the specified time (usually just a few minutes) deactivates this enzyme. And here I thought it was just to bring out the pretty green color of my Romano beans. Blush.

Up until about ten or twelve years ago, I had never used a pressure canner. Pickles and jams and most fruits and tomatoes are classed as high-acid foods, and as such they are safe to process in a boiling-water bath. Vegetable like green beans and beets, however, unless they are pickled, are low-acid foods. Meats and foods containing meat (such as chili or spaghetti sauce) are also low-acid foods, and must be processed in a pressure canner.

This point cannot be over-emphasized. Get yourself a good up-to-date canning book (see the recommendations in Appendix B); it will have charts that show clearly which foods may be safely processed in a boiling-water bath and which require the use of a pressure canner.

Over the years, and also during the Master Food Preserver course, I have heard many stories, horror stories actually, about frightening mishaps with pressure canners. Almost everyone I've ever talked to about pressure canners seems somewhat afraid to use them. Considering these stories, I can understand this. At the same time, it is such an essential tool in my food-preservation arsenal that I want to encourage you to read on, and keep an open mind on the subject.

I am fairly certain that virtually all of the pressure canner accidents and mishaps I've ever heard about were caused by operator error. Because of the buildup of pressure and high temperatures (at ten pounds pressure, it is 240°F), there is a danger of potentially severe injury from burns or flying objects if a pressure canner is used incorrectly. I have to admit to being very nervous the first several times I used my new pressure canner. I sat in a chair at what I hoped was a safe distance away, but close enough to see the pressure gauge. User's manual in hand, I read and reread the instructions, looking up at the pressure canner every couple of minutes, going over in my head the sequence of events:

Jars in the canner, lids on with rings snug but not tight, three quarts of hot water. Check to make sure the lid is locked in place. Bring it to a boil, and when the steam is coming rapidly out of the steam vent, set the timer for ten minutes. When ten minutes are up, put the weight on the steam vent. Now keep an eye on the pressure gauge, since the pressure is going to start to build.

Once the required pressure is reached, I set the timer for the length of time required for processing. At this point, I need to start gradually turning down the heat, or the pressure will keep going up. I use a gas stove, so it's easy to gradually turn down the heat. I learned the hard way not to turn it down from high to low all at once; if the pressure drops below the required level, you will have to turn it back up and start timing all over again once the correct pressure is reached.

Compared to using a boiling-water bath, processing times with a pressure canner can be quite lengthy. For example, when I process cooked chicken meat, pint jars require seventy-five minutes. Chicken stock is faster, processing for thirty minutes for quarts.

I always have to remind myself that processing times must be adjusted when you are more than a thousand feet above sea level because the boiling point of water goes down as pressure increases with elevation. My kitchen is at almost exactly a thousand feet. My book says to add five minutes of processing time between one thousand and three thousand feet. Sure, I could probably get away with skipping the extra time; for all I know, the top of my stove is a little below a thousand feet. But we're talking food safety here, so why take the chance?

Please note that this description above of my pressure-canning process is simply a brief rundown of the steps involved. It is not intended as a substitute for reading and following the instructions with your pressure canner! Please, read and save those instructions and refer to them every time you use it to process food.

I don't know about you, but I use my pressure canner year-round. You'd think I could skip looking at the instructions; after all, I must know them by heart by now. Well, to some extent this is true, but I have long since been in the habit of looking up the processing times for chicken or chili or whatever, every single time. It is too easy to get confused between processing times for pints or quarts, or between raw-pack and hot-pack or whatever. It only takes a minute to check the charts, and I have the peace of mind of knowing that I have done all I can to ensure the safety of my processed food. Also, going over and over the instructions will increase your self-confidence, not something to be overlooked when using a potentially dangerous piece of equipment.

Wondering why I don't simply freeze my chicken or turkey or chicken stock? Well, first of all, we don't have a large freezer. Our small propane refrigerator/freezer has a total of eight cubic feet of space; about one cubic foot is the freezer. Propane freezers, like refrigerators, are much more expensive than electric ones, and honestly, I don't know where I would put a chest freezer anyway. I also heard plenty of stories in the master food preserver class about food lost in the depths of chest freezers, only to be found a year or two or three later and thrown away.

For me, though, the best thing about canning as opposed to freezing food is that I never have to thaw anything before using. It's so quick and easy to grab a jar of minestrone or spaghetti sauce or smoked tuna off the shelf, pop the lid and heat it up. Especially during the short days of winter, when daylight time is at a premium for getting outdoor chores done, it helps greatly to be able to put a delicious hot meal on the table in just a few minutes.

CHAPTER 29

Egg Money

J UST IN CASE YOU HADN'T NOTICED, *Pure Poultry* is not about how
to run a major income-producing business. I hope you're not disappointed. However, it offers a few ideas to start you thinking about
how to bring in a little extra cash while providing a valuable service
to your community.

Don't assume that, just because you live in a rural or farming area, the
market for farm-fresh eggs or chicken is saturated. That's what I thought,
before we decided to apply for an egg dealer's license (required in
Washington State to sell eggs wholesale; see 'WSDA' under 'Businesses
and Organizations' in Appendix B). Our area has a rich farming history.
In 2007, the first year we had chickens, I remember driving around the
town and frequently seeing signs at the end of driveways advertising
fresh eggs for sale. Although we knew our eggs were organic, super-fresh
and delicious, I resisted the idea of selling them at first because I couldn't
imagine how we could compete with all the other established egg
producers near us. It never occurred to me then to look into the possibility of selling the eggs wholesale, for instance to a restaurant.

In the spring of 2008, we participated in a sustainable agriculture
class put on by our local county extension office. Boy, did we learn a
lot during those few months. It was mainly about how to start and
grow a successful farm enterprise, and provided us with some valuable resources of information, including *The Green Book*.

The Green Book, a handy booklet compiled by the Washington State Department of Agriculture, is updated every few years. It offers comprehensive advice and specific information about licensing and other regulations pertinent to selling and processing farm products. I remember reading it from cover to cover, and in the process, I came to the section on, you guessed it, eggs.

I soon discovered that getting an egg dealer's license is neither complicated nor expensive. While in the sustainable agriculture class, I requested and submitted our application.

In about two weeks, we received our egg dealer's license, almost eleven months after we got our first chickens.

In the meantime, we had struck up a friendship with Gabriel and Jessica Schuenemann, the young couple who own the Alder Wood Bistro in Sequim. They are committed to sourcing ingredients locally whenever possible, as well as using organic foods. Although the Bistro does not serve breakfast, Gabriel told us that they were using about fifteen dozen eggs every two weeks, mainly in their house-baked desserts such as crème brûlée, supplied by a large distributor. At a point where our license application was in the mail, we invited Gabriel and Jessica over to dinner. I know we served them two or more dishes featuring our eggs, but I can only remember the Mexican-style flan. (I also recall asking Gabriel if it was weird to have someone else cook for him. He said, "No, actually, it's kind of nice.") We sent them home that night with a dozen fresh eggs.

The next day, Gabriel called up and said, "Your eggs are delicious. We'd like to use them at the Bistro." I don't recall whether we had already told him that we had applied for the egg dealer's license, but I assured him that it was on its way and we would love to sell our eggs to him once we had the license in hand.

As I write this, in early February 2013, the Alder Wood Bistro has been buying our eggs for nearly five years. In early 2009, when our first batch of ducks were laying consistently, they began buying our duck eggs as well as chicken eggs.

As I said before, don't take it for granted that there isn't room for your farm-fresh poultry product in the local market. Do your research. Think about what you can do to create your own niche market by emphasizing some unique or unusual aspect of your farm or products. Do you free-range your poultry? Is your operation certified organic? Do you grow only heritage breeds? Invite neighbors and potential customers to your farm and don't be shy about explaining to them why your eggs or turkeys are better than the other choices locally available. You might want to have a nutritional analysis done on your eggs and educate your customers about the health benefits of free-range or pastured eggs and poultry.

If you have potential wholesale customers such as restaurants or grocery stores, be generous in giving out free samples of your products. Don't assume that experienced chefs know the difference between commercial eggs and poultry and your fresher, locally grown products. In the case of the Alder Wood Bistro, chef Gabriel was not aware that buying eggs from a local producer was as simple as that producer obtaining an egg dealer's license. It would be worth contacting your state's Department of Agriculture to find out what the regulations are in your area.

A tricky aspect of selling to restaurants is that their needs tend to fluctuate seasonally. The Bistro is busiest in the summer months, especially during the week of Sequim's Lavender Festival in July. Our egg production is usually highest in mid- to late spring. We try to keep our laying flock to a size that will produce enough for the Bistro's busiest months, but what about the rest of the year? In early 2012, when our laying flocks of ducks and chickens were the largest they'd ever been, we found ourselves with many extra eggs even after we had supplied the Bistro. Because we felt part of the appeal of our eggs was the freshness, we needed to find someone, and quickly, to buy the extra eggs.

It turned out to be easier than I'd expected. I got on the telephone, and by the end of that same day, I had orders from two local food stores for duck eggs. It is certainly a little more work to supply more than one customer, what with deliveries and billing and all that, but that spring we were able to sell virtually all of our eggs. None of them went to waste, and some happy customers discovered that, even with eggs, locally grown is better.

We have never sold our eggs or poultry at farmers markets, but this is an increasingly common option in many urban and rural communities. Your state will have a farmers market association; look it up on the Internet or ask at your local farmers market. They will be able to tell you about what you can sell there, labeling requirements, benefits of joining the association and more. (In our case, *The Green Book* has a section on farmers markets, and includes handy information like telephone numbers and web addresses.)

As far as I'm aware, as of January 2013, we were the only licensed egg dealers on the Olympic Peninsula who sell duck eggs. Once people around here figured out that there were duck eggs available, the demand has consistently outrun the supply.

If you breed your poultry, selling hatching eggs and/or chicks or poults (baby turkeys) is another possible means of earning some money with your heritage birds. However, please be aware that breeding and hatching quality poultry can be a time-consuming and complicated task involving good management practices and record-keeping. I strongly recommend that if you do plan to breed your birds, limit yourself, at first anyway, to one breed of bird. It's much easier if you don't have to isolate breeds from each other, and you'll always know you won't be stuck with any cross breeds.

How about a U-Pluck operation? I mentioned that, when we slaughter turkeys, customers often ask to participate in the process. Sometimes this is because they are considering starting to raise turkeys themselves; others simply want to be involved in whatever way they can with the production of their food. In Washington, many counties have some sort of group or club that pools its resources to buy scalding and plucking equipment that members can borrow or rent to process poultry. If you have the means, and you have found that there is interest in your community, you can start your own group or simply let people know when you will be slaughtering your own birds and invite them to bring theirs and have a big slaughtering party. Trust me, they will gladly pay a reasonable fee to use your equipment, and they will be so happy to have someone with experience walk them through the process.

Egg Money

My mother, Susan Redhed, grew up in a farming area north of Chicago, Illinois. Although she was more interested in raising hogs than poultry, she remembers chickens being part of family life after moving to their farm in the late 1940s.

"That whole area was farms when we bought ours," she recalled, "interrupted only by tiny lake communities like Forest Lake, our previous home, or towns like Lake Zurich (population 300), where we went to school. My impression was that most farms had chickens.

"Dad had been in the National Guard for some time before World War II, and also apprenticed as a carpenter. He was building houses in Lake County when the war broke out. He tried to join the Army, but was rejected because of his poor vision. So he took a job as postmaster of Prairie View, a tiny railroad town. The post office, which was probably a converted small house, had a backyard with a small shed in it.

"Of course, this was before we bought the farm. Farming was in Dad's blood. His family had owned a small farm for several years, and he loved working for his uncle Louis Klein as a boy. He must have always dreamed of having a farm of his own. So this was a start: He bought a few hens, established them in their new shed and waited for some eggs.

"I'm sure we got a batch of chicks, New Hampshires, not long after we moved to the farm, and we had them steadily after that. We kept them in a fair-sized wooden outbuilding, adjoining a small yard fenced in with chicken wire. Occasionally we let them out to run and grub around the barnyard. We generally had at least one horse, a cow and my pigs. We had four Hereford steers for quite a while. And of course Pet, the cow, produced a calf every year. All the animals were kept separated from the chickens.

"TNT, the skinny little cock, was named for his habit of racing all over when we let him loose in the barnyard, and being something of a pest to his flock-mates. Oh, yes, I remember Dad putting glass or wooden eggs in the nest boxes to give the girls the idea.

"We kept chickens mainly for food, I suppose, but I remember Mom talking about her egg money, which was typical of farm wives in those days. As far as

I remember, Mom was the only one in the family who sold eggs, probably to friends at Forest Lake or others nearby who didn't have chickens of their own.

"We must have always had some cocks around, because we certainly had chickens to eat regularly. Dad just pulled out his hatchet and sacrificed them on a handy stump."

Chapter 30

The Bigger Picture: Poultry in the Community

O NE OF THE MANY INTERESTING THINGS WE'VE LEARNED on the farm seems paradoxical: The more steps we take toward self-sufficiency and responsible stewardship of our land, the more important it is to be connected with our friends in the community around us. There is this sense that each of us has something to contribute, whether it's time, money, skills, equipment or connections. Then there are the less tangible things, such as leadership qualities, even enthusiasm and passion. This may seem counter-intuitive to those of us who are accustomed to a system where competition determines who "wins." Yet this kind of cooperation — dare I suggest, interdependence? — has historically been a major factor of rural life in general, and farming communities in particular. For us, part of the idea of sustainable farming — and living in general — is that by sharing the resources we have with others who share our values and goals, we can all ultimately be successful in our separate endeavors.

In the class we took in 2008, we learned that one of the underlying principles of sustainable agriculture is the social or community aspect. Of course, there are many possible ways to involved in the community. Once again, it starts with planning (see Chapter 6). Do you have kids? Might they be interested in raising a rare heritage-breed chicken as a 4-H project? Maybe your town has a farmers market. Perhaps by selling a few eggs or jars of honey or bunches of kale every weekend (bring the kids to help), you'll discover new friends, earn extra cash and have the satisfaction of sharing the abundance of your flock or herd or garden.

You may reasonably ask, what about those who don't live in rural areas? Perhaps you live in a city apartment or a suburban home with a small backyard and neighbors close by. Maybe you have a large playful dog, or small children or two jobs. How can you enjoy the benefits of heritage poultry? You guessed it: By getting involved in your community.

If your town doesn't allow chickens or, for whatever reason, you don't think it would work for you, start asking around. Somewhere among your neighbors, friends, co-workers or church group, you will either find or hear about someone who has poultry.

Be creative! Maybe you want to start with a couple of dozen fresh eggs per week. Well, what can you offer in exchange? Of course, you can simply pay cash, but believe me, farmer types also love to barter. And everyone has something of value to share. Even you.

Suppose you don't even have a backyard. How about getting together with a few friends or neighbors to form a small co-op? Or negotiate with a neighbor to share some garden or coop space. Handy with tools? Why not trade some coop-building or fence-repair labor for a regular supply of fresh eggs?

In recent years, the popularity of keeping backyard chickens has increased so much that many large cities and incorporated areas allow residents to raise chickens. Other cities have changed the existing laws to permit more birds. A few years back, for example, Seattle increased the number of chickens allowed in the city limits from three to eight.

Chickens in the City

Town dwellers love fresh humanely raised chicken and eggs, too. Why not start an Adopt-a-Chicken program? The customer can chip in for the birds' feed and bedding, and maybe even help out with the chores occasionally. You'll have to figure out what the eggs or stewing hens or roasters are worth, and negotiate the details with your customers. Trust me, it's not that much work, and it's a terrific way to make new friends and establish a presence in your neighborhood. At the same time, you will be educating your customers about heritage poultry and why sustainably produced eggs and meat are better than the mass-produced variety.

Mentors

It's a good idea, especially if you're new to raising poultry, to try to connect with someone with more experience. Ask if he or she would be willing to act as your mentor. This can be so helpful when you have a concern or question about your birds' health, feed, predators or whatever. If you have experience raising poultry, especially the more unusual breeds of chickens, guinea fowl, waterfowl or turkeys, I'd encourage you to look for ways to share what you've learned. It doesn't have to take a lot of time to be a mentor. You will find that you learn in the process, too; it's definitely a win-win situation for everyone involved, including the birds.

Community Canning Kitchens

I'm very interested in the idea of the community canning kitchen. Some of you may remember these kitchens, if you were growing up during the war years of the early 1940s. A local commercial kitchen, often at a church or perhaps a county grange, opened its doors to everyone needing help putting up their harvested fruits and vegetables and meat. The neighborhood housewives, along with their older children and probably some men too (the ones who weren't away at war), would get together, bringing their homegrown produce and perhaps a freshly slaughtered chicken or lamb. The kitchen usually had at least one person on hand who was an acknowledged expert in the arts of food preservation. This person was available to answer questions, give instruction in using pressure canners and other kitchen equipment and walk everyone through the process of putting up their food. Once all the meat, vegetables and fruits were in jars, everyone was free to swap their preserved food with others before heading back home at the end of a long day.

The community canning kitchen must have been a godsend for the women, men and children who were obliged to do the best they could to keep their households running under difficult circumstances. Many participants lived on farms or had market gardens and some livestock. It must have been challenging just to find the time to harvest the crops when they were ready, much less process them for long-term storage. And with rationing and other wartime realities that made frugality a true necessity, no one could afford to waste anything. The community canning kitchen was a place to spend the day among friends who were experiencing the same difficulties, to lend a generous hand and receive willing help from others and turn a necessary but stressful task into something much more pleasant and satisfying.

I would love to see a revival of this concept; it would be possible virtually anywhere, in town, the suburbs or the country. Many churches, schools and community centers have kitchens. Maybe a local restaurant would allow the use of its kitchen on a day that it's closed. Of course, it could also be done on a smaller scale; for instance, it could just be three or four neighbors meeting in one of their homes to can tomatoes when someone has a bumper crop. Or, if you own or have access to a mechanical plucker, you might organize a slaughter-day work party, inviting others you know who raise poultry for meat. Believe me, no matter how many or how few chickens you are slaughtering, having one or two extra pairs of hands to help can make a big difference. It's more fun, too.

I'm sure you will think of other useful ways to get involved in your community. I could go on (and obviously do) with my own brainstorming and toss a few more ideas out there. But since your situation is unique, I'll leave you to get your own creative juices flowing and think about what you could use in the way of help or encouragement, and also what you might have to offer. Talk to your spouse or partner, and your kids — even young kids love to be included in this kind of thing. And the whole community will ultimately benefit from your efforts to stimulate such helpful interaction and cooperation.

CHAPTER 31

Chicken Coop for the Winter-hardy Soul

B RRR ... 35°F HERE AS I TYPE AT 4:15 ON A SEPTEMBER MORNING. Actually I look forward to this time of year; fall has always been my favorite season, and I love writing here in the living room next to the wood stove. I wonder what kind of winter we'll have? Although there was plenty of cold weather, last winter we had very little snow. I don't worry any more about how the birds will handle the cold; as David says, that's why they wear those nice little down jackets.

The birds are nearing the end of the moulting season, showing off their perfect new feathers as they sunbathe on a warm late-September afternoon. Meanwhile, the trees around the farm are already starting to shed their fashionable summer looks. The leaves of the vine maples are rapidly turning shades of gold, and in the slightest breeze, the black cottonwood leaves flash their silver undersides like Victorian debutantes flirting at a ball. October rains are just around the corner, and I'm thankful that this year's early moult has supplied the birds with their new down coats before the winter chill takes hold.

I remember feeling confused, though, when first researching chicken breeds, trying to choose types that would do well up here in the mountains. All the charts said "cold-hardy," or "not very cold-hardy," but they didn't explain exactly what that meant. Would we need to heat the coop somehow when the temperature dropped

Nankin rooster in snow.

below, say, 20°F? Surprisingly, I couldn't find any information or suggestions that were any more specific.

So how do we keep the birds warm through the dark cold winter months? I was surprised to learn that, for most breeds of chicken, you don't need to be too concerned until the temperatures get down below 0°F. Cold-hardiness, as mentioned earlier, was one of the criteria I used for choosing what breeds of poultry might work well in our climate.

The only exception we've made in breed choices as far as cold-hardiness goes is the tiny Nankin bantam. They are not known to be exceptionally cold-tolerant. However, they have done remarkably well here, sleeping in unheated, uninsulated coops like all of our poultry. Some Nankins have a rose comb, but ours are the single-comb variety. The cold doesn't ever seem to bother them a bit.

One of the main cold-weather issues for birds is frostbite, which most commonly affects the combs of roosters, especially single-comb types. Roosters' combs are usually larger than those of hens. Also — another surprise to me — roosters don't tuck their heads under their wings when they sleep, like the apparently smarter hens do, so their combs are exposed to the air and more vulnerable to frostbite. Fortunately, we've had no problems with that, although the past two winters we've had our share of single-digit temperatures.

I've heard that rubbing petroleum jelly (Vaseline) into the combs of roosters will help prevent frostbite. Our wintertime temperatures are often in the teens, sometimes in single digits. To the best of my recollection, they have not gone below 0°F since we've had poultry. We've never seen the slightest hint of frostbitten combs, so we haven't tried the Vaseline method. If your temperatures regularly go below 0°F, though, talk to other poultry growers in your area, or perhaps your county extension agent, and ask for recommendations.

There's a Chill in the Air

OK, so it's not surprising that October chill is in the air, considering that it is, in fact, October. Yesterday morning, we had our first frost: 31°F. Around October 10 has been average for this event, although we were surprised two years ago when it came three weeks early (to the extreme detriment of my later bush bean crop). With a low of 34°F this morning, it was frosty down in the lower pasture where the pigs currently reside. It's gorgeous and sunny now, though. I'm going to finish up some new nest boxes and put a roost in one of the coops for the young Nankins; they're almost five weeks old now and more than ready to start roosting like the big kids.

As you know, we live off the grid. This does make some difference when it comes to caring for the animals over the winter, but not as much as we had anticipated. The main issue is to keep the water in the drinkers from freezing. Lots of people use heated (electric) waterers. Frankly, even if we had that option, it would be problematic just because of the area that the birds range on; the feeding stations are quite spread out. We simply add warm water to the drinkers first thing in the morning if they're icy; we check them frequently through the day, positioning them in the sun whenever possible. Yes, it's a little extra work, but we're out there checking on the birds regularly anyway, which is always a good thing to do.

The Importance of Roost Size

Probably the most important fact I learned (eventually) is how chickens roost; surprisingly, the width or diameter of the roost itself makes a difference. It needs to be large enough that the birds' feet don't wrap all the way around. It's hard to imagine a roost being too big, but it can certainly be too small. When chickens settle down on the roost, their feathers cover their feet, and if their little toes go all the way under the roost, they won't be under that toasty down blanket. If it's cold enough, this can cause frostbite. So, for chickens, we use nothing smaller than two-by-twos for roosts; for turkeys, a two-by-four with the wide side up seems to work well.

Make sure that there is also plenty of roost space for your birds. Take into consideration not just the number of birds you have, but also their breed and size. You'd be amazed how common it is to assume that turkeys don't need more sleeping room than chickens, but believe me, they do.

Another Tip for Keeping Birds Warm at Night

One trick to help keep your birds warm at night is to feed some cracked corn about an hour before they head into the coops for the night. The extra carbohydrate will give their metabolism a boost, generating more body heat during the long winter night.

A strange thing happened yesterday afternoon. One of the Cochin banties was sitting on the old split-rail fence, wheezing, sneezing and generally sounding pretty terrible. After chasing her around for a while, David caught her and brought her inside. She was gasping and didn't seem interested in eating or drinking anything. We decided to keep her warm and dry and see what happened. Just a few hours later, she was quiet, not wheezing or making that painful-sounding noise. We kept her inside overnight, and she was quiet all night, then seemed pretty perky this morning. David let her out the back door, and she seems to be OK so far. We're guessing she either ate something that slightly poisoned her or possibly had some kind of allergic reaction. Very strange.

This is a good example of why you should regularly walk among your birds, looking for little signs of trouble like coughing, sneezing, odd behavior, abnormal-looking poop, that kind of thing. With the number of birds we have, it would be pretty easy for something to spread around the flock if we didn't catch it early enough. It's also a good reminder that you can easily help to ensure your flock's health by making sure their feeders and drinkers are kept clean.

It has occurred to me that living off the grid has probably motivated us to be a bit more creative in some ways; for example, not using heated drinkers that require electricity. Of course there's nothing wrong with heated drinkers; we've simply found that there are easy and cheap alternatives to many things that we might take for granted if we had full-time electricity. And hey, I actually enjoy cutting firewood to heat the house! Ours, I mean.

CHAPTER 32

Back to Standard Time,
Which the Birds Never Left

O H, GOODY. We've "fallen back" to standard time once again. Those of you who are familiar with my collection of various soap boxes will appreciate the fact that I can reasonably get on this one just twice a year. Honestly, now, don't you think "daylight savings" loses some of its supposed meaning when it lasts for more than seven months?

If you have poultry, specifically one or more roosters, you know how the change to standard time has affected them: *not at all*. Our roosters, bless their little hearts, continue their routine of crowing for a few minutes about an hour before dawn. Do they care that our atomic clocks reset themselves at 2 AM last Saturday? They do not. Did they gather around the three-gallon drinkers to discuss how to use the "extra hour of sleep?" Doubt it. I love it that these supposedly unintelligent creatures make this transition so smoothly, and do you know why they do? Because nothing *actually* happened!

For many years now, I have not worn a watch, so it's interesting how much I am aware of the time. And although the birds are blissfully unaware (I think) of what the hour is at any given time, they do definitely have their routines, which probably accounts for why we're paying attention to the time during the day! As sunrise approaches, the roosters start crowing. (It's a myth, though, that roosters crow simply to announce the arrival of dawn; if they do happen to crow right at sunrise, it's frankly just a coincidence.) Once it's fairly light, all the birds become active and want to leave their respective roosts and head out for a busy day of foraging, dust-bathing, sunbathing, mating and debating the finer points of the pecking-order rankings.

The ducks, who seem to stay up late at night partying (we can hear them talking to each other at all hours), are active and energetic for the first couple of hours in the morning, then settle into a nice long nap. Lately they haven't been going down the hill to the pond, but when they do, they always run back up once or twice during the day to get a snack; they are amazingly consistent about the timing of this.

The turkeys also have regular nap times, but mostly their routine follows ours: When we're outside, they follow us around. When we're inside, they walk around the house, peering in every window, trying to see where we are and what we're up to. (Makes me glad our bedroom is on the second floor.) On sunny dry days, they like to dust-bathe with the chickens, always quite a sight to see. Birds synthesize vitamin D from sunshine like we do, and the turkeys and chickens look funny when sunbathing; they recline in a sprawling wings-outstretched position that sometimes makes them look, well, dead.

Our glider is a favorite daytime roost for young turkeys.

Toward the end of the day, about an hour before heading into their coops, all the birds gather around the feeding stations for a bedtime snack and a nice drink of water. This time of year, we generally start giving them extra corn before we tuck them in at night; the additional carbohydrates help them generate body heat while they sleep.

What it comes down to is that our chickens, turkeys and ducks are not enslaved by any clock but their internal ones. They get up when it gets light; they rest when it gets dark. It's true that we humans have to find a balance between being completely schedule-driven and being completely selfish about how we use our time. However, I do think it's good to remember that it wasn't so long ago that we didn't have electric lights to artificially extend our waking hours. The birds know how to make the most of the available daylight hours at any time of year. Maybe they know something we don't.

CHAPTER 33

Warmth

O NE OF THE OBVIOUS ADVANTAGES of living on a wooded property is the natural abundance of firewood. Of course, much has
to happen between the woods and the wood stove. Felling trees is
extremely dangerous work — I've heard it compared to working with
explosives — and fortunately David is both skilled and experienced
in this task. He often has me come out with him when he is about to
cut down a tree. A second person on hand helps, especially if the tree
in question is a large one; often we are dropping trees over a hundred
feet tall. It is virtually impossible, when you're standing at the base
of a tree of this height, to determine when the top of the tree has
started to move. Also, you never know when a sudden gust of wind
will come up and set the tree swaying when you're not expecting it.

David always makes sure that we both have at least one escape
route planned before the cutting begins. Sometimes the tree simply
doesn't fall in the direction you planned; fortunately, largely due to
David's skill, this doesn't happen often. In addition, when the tree
does fall, it usually knocks branches off adjacent trees. Hence the need
for hard hats, even for the person who isn't actually doing the cutting.

Once the tree is down, we knock off the smaller branches with the
limber (a fairly lightweight double-bitted axe) and thicker branches
with the chainsaw. The larger branches are usually sawed up for firewood too.

Next the log is cut into lengths that work in our wood stoves. The rounds are hauled to the woodshed near the house. If the tree is near a road in the woods, we can use the pickup; otherwise we use a wheelbarrow, or even a sled when there's enough snow. Finally, the rounds are split and stacked in the woodshed according to type of wood (we have several kinds of fir, and red alder) and whether it is green or dry. Our woodshed is capable of holding about twelve cords of split firewood, but as we're heating our house for seven or eight months of the year, it seldom stays more than half full for long.

Sure, it's a lot of work. We always hope to get plenty of wood split and into the shed in the spring, but that's difficult to accomplish. There are just too many other things going on this time of year. Egg production is high, so more time is spent processing, selling and delivering eggs. It's the busiest season for gardening, since the bulk of the planting happens in spring. And poultry breeding is in full swing as well. So realistically, it's usually late summer before we're able to put some serious time into the gathering of firewood.

Personally, I love splitting firewood, especially since David bought me my very own six-pound splitting maul. (Nothing says "Happy Birthday" like a splitting maul, you know.) It has enough weight that a person of my phenomenal strength can use it with some authority, while not being so heavy that it wears me out in the first five minutes. I can generally swing away for an hour or more, depending on the type of wood and how dry it is, before needing a break. So usually, David hauls the rounds out of the woods, and I do most of the splitting and stacking.

We have two wood stoves: one in the living room and one in the kitchen. Together they do a great job of heating the house. The kitchen stove is a beautiful wood-burning cook stove, complete with a roomy oven and a five-gallon copper-lined hot water reservoir with spigot. We keep large pans of water on this stove during the winter. This helps humidify the air (wood heat does dry out the air), and also to provide extra hot water for dishwashing. Yes, we do have a gas hot-water heater, but it's nice to save a little gas when we can. I also like to keep a pot of soup or stew or chili simmering on this stove too; it's so handy during the short winter days when it seems like there is hardly time to eat, much less cook.

The living room stove is a soapstone and cast iron model (see Appendix B). It features double walls of soapstone, with an air space in between. The soapstone efficiently absorbs and radiates heat, even long after the stove has gone out. A heat-powered fan on the top helps direct the rising heat into the room. A shallow pan of water placed in front of the fan improves the humidity.

Our gas wall lamps also contribute some heat; David says each lamp is the equivalent of a 400-watt heater. All I know is that, between the solid construction of the house, the wood stoves and the gas lamps, we stay comfortably warm even in the coldest months.

CHAPTER 34

Pure Poultry Premise #1:
Purebred Birds Are More Sustainable

I N THE PROCESS OF RESEARCHING AND WRITING THIS BOOK, I have realized that it is a genuine reflection of our values and the kind of lifestyle we choose and strive to achieve. In fact, unlike most currently available books about poultry, it is less a "how-to" than a "why-to" kind of book. Throughout, I have endeavored to share the choices we have made — and the reasons for them — in our progress toward a more sustainable way of life.

What does "sustainable" mean? I recently reread Michael Pollan's *The Omnivore's Dilemma: A Natural History of Four Meals* (see Appendix B), and I agree with his view that the term has in recent years been overused to the point where, for many people, it has little meaning. He also suggests that anything unsustainable "sooner or later must collapse." For us, so far, "sustainable" has the most relevance in regard to our animals.

To summarize, then, we have been growing heritage-breed chickens, turkeys and ducks. Over the past few years, we have raised a couple of pigs for about half the year. The chickens and ducks are mainly kept for eggs; we keep a small breeding flock of turkeys and raise up to two dozen every year to slaughter. The pigs, while grown for meat, also provide a valuable service by plowing up previously unused fields and eating the roots, allowing us to reseed with grasses and clovers to create better pasture. Choosing the breeds we raise involved much research and a conscious decision to keep only pure-bred animals, also known as "heritage" breeds. From all we've read and heard, as well as learned from our own experience, it is clear to us that purebred livestock is the best choice when the aim is greater sustainability.

For example, as mentioned in an earlier chapter, the Broad Breasted turkey (a commercial hybrid) cannot mate naturally; it must be artificially inseminated. Cornish Cross chickens, the ubiquitous grocery-store birds, have been bred to reach an impressive broiler size in seven weeks or less. Obviously, these chickens are being slaughtered long before the normal breeding age of eighteen to twenty weeks (depending on variety). However, their warp-speed growth comes at a high cost: leg problems and sudden death from cardiac arrest are not uncommon.

I have no quibble with those who are trying to make money in the poultry business and have found that they must raise these fast-growing hybrids in order to stay afloat financially. After all, no business is ultimately sustainable if it loses money. Also, we don't raise chickens specifically for meat, although we do occasionally slaughter a few roosters when we have more than we need for breeding. So perhaps my view doesn't seem to carry much weight. What I've said in the previous paragraph, however, are known facts, not just my opinion. That said, I hope you don't hear any judgment there; I assure you I feel none.

Aside from reproductive issues, another reason we prefer purebreds is the variety of instinctive behaviors that contribute to the long-term health of the farm. For example, one of my criteria for choosing a chicken or turkey breed is foraging ability. All our birds free-range during the day, and we expect them to do some of the work of feeding themselves. (Statistics claim that foraging can account for up to 30 percent of a chicken's daily feed requirement, but I haven't seen any data as to which breeds were sampled or what kind of environment they had to forage in.) Our observation has been that the birds spend a few minutes eating grain first thing in the morning, then head off to happily scratch and peck until it's time for a late snack before heading in for the night.

Our Midget White turkeys also are excellent foragers. The first year we had them, they did a great job of cleaning up the windfall apples near the house, which kept the deer from coming into the yard. The ducks love to dabble in the freshly plowed fields after we move the pigs to another paddock; they break up the manure piles and eat the larvae of intestinal worms, thereby disrupting the cycle of parasites and disease. The heritage pigs have thankfully retained their instinctive rooting behavior; they find a good portion of their own food and efficiently rid the land of the roots of weeds and other unwanted plants. The chickens and turkeys then follow the ducks, picking out leftover weed seeds and worms. This progression, by the way, also closely imitates nature; for example, woodland birds break up bear manure, in the process spreading around seeds from the berries the bears eat in large quantities in the summer.

We continue learning about how to fully utilize purebred livestock on our farm. In 2008, when we participated in that sustainable agriculture class, one of the ideas that impressed us was that the key to success on small farms is diversity. We have already had a taste of the value of diversifying, step by slow step. In the next few chapters, I will further detail how heritage poultry contributes to sustainable agriculture and how we are working toward our goals.

CHAPTER 35

Pure Poultry Premise #2:
Shorten the Food Chain

WE'VE ALL HEARD A GREAT DEAL RECENTLY ABOUT "eating locally"; that is, trying to be more aware of exactly where our food is coming from and choosing to buy locally produced ingredients for our meals whenever possible. More and more restaurants, such as the Alder Wood Bistro here in Sequim, have enthusiastically embraced this idea as well. More consumers are also realizing the benefits of locally grown produce and meat, even visiting farms and farmers markets and buying directly from the producer whenever possible.

What are these benefits? Well, obviously when you can buy directly from a farmer, the food is sure to be fresh and in season. Also, locally grown food, especially produce, is usually bought when it is seasonally available, which translates not only into better quality and flavor, but also lower prices. Fruits and vegetables bought out of season most often come from another state or even another country, and you can be sure the cost of all that transportation and cold storage is being passed on to you. In addition, if you've ever visited a local farm and bought directly from a farmer, you will have realized the difference it makes in the way you relate to what you eat. So much of our culture of eating involves relationships, and buying locally produced food promotes beneficial relationships, not just with an individual producer, but with the community at large. Which brings up yet another benefit: keeping our money in the local economy.

What's happening here is we are shortening the food chain, the often lengthy and convoluted path our food travels to become part of our next meal. Between the lines of the benefits outlined above are the less obvious costs, to us as consumers and to our environment (and even our society) as well. A few statistics: On average, seven to ten calories of fossil fuel energy are used in the process of delivering every single calorie of food energy to your plate. Twenty percent of the total energy used in food production is consumed on the farm; the rest is spent processing the food and moving it around. (Growing organically uses about one-third less fossil fuel than growing conventionally, but that savings quickly disappears if compost used on the farm is not produced either onsite or nearby.)

By contrast, buying even part of our food from local sources can make a big difference in the true cost of feeding ourselves. For example, if you buy a holiday turkey from us, the food chain is very short: We grow the turkey and slaughter it, and you pick it up within a day or so of slaughter. It isn't even frozen. Some customers have come up to help on slaughtering days, adding another level to their participation in their personal food chain.

Restaurant owners like Gabriel and Jessica Schuenemann of the Alder Wood Bistro are also helping to set a new standard in our community. They are committed to sourcing locally as much as possible of the food they serve, which is great for small farms like ours. They have been buying organic chicken and duck eggs from us since 2008, delivered several times a week, so the chain, once again, is short: From farm to restaurant to consumer's plate, the eggs have traveled less than eight miles. By way of contrast, recent studies estimate that any given ingredient in a typical American meal has traveled an average of 1,500 miles by the time it arrives, no doubt exhausted and disillusioned, on your plate.

Our farm is small. We don't have the room to sustainably produce more pork or poultry than we do now, since we are unwilling to compromise our standards of allowing the birds to range freely during the day. It has occurred to us that, for a production level like ours, our relationship with the Alder Wood Bistro actually is a model for how small farms can be successful. So much of our national food production (and yes, even organic food production) is done on a huge scale, at a huge cost on a number of levels, many of which are hidden. What if more small producers sought out relationships with local restaurants, or even small community grocery stores, to help supply the growing number of consumers enthusiastically seeking to connect with these producers?

After all, for consumers to buy locally produced food, someone has to produce and sell it. Removing a number of links in the typical American food chain by selling either directly to the consumer from the farm or through a forward-thinking restaurant just down the hill has become a crucial piece of the big picture for us. It is also an important part of our concept of sustainable living.

CHAPTER 36

Pure Poultry Premise #3:
Challenge the "Get Big or Get Out" Adage

A FEW YEARS AGO, I was asked if it's possible to actually make any money raising poultry. My somewhat facetious response was that we'd find out after my book about raising poultry was published.

Seriously, though, the query did get me thinking at a deeper level about what it is that we're doing here. Are we trying to make a living or make a life? Be more self-sufficient or more involved in our community? Indulge a hobby or create a lifestyle that reflects our values, our spirituality, our worldview, even our politics? I am quickly arriving at the intriguing conclusion that, to one degree or another, it's all of these things.

Those of you who know me are aware of my love for the "what if" questions. Here's one to consider: What if the "get big or get out" idea of farming is simply *wrong*?

"Small" farmers, you know what I'm talking about. Supposedly wiser minds than ours want us to believe that, short of buying up more and more acreage to raise the same crop (read "commodity") year after year, it's just not possible to make a living in today's farming world; hence, "get big or get out." Never mind the costs: the depletion of soil elements critical to its long-term health, the need to use more and more chemicals over time, the financial stress of increasing debt and uncertainty. Funny thing, the marketing gurus behind such ideas don't mention these costs; perhaps they aren't aware of them, or maybe they just don't care. After all, they're in marketing, not farming.

Here's another question: What if "get big or get out" aren't really the only choices? To me, this mantra implies that, at the end of the day, what's important is the profit margin. In other words, the bottom line is business. I think it's time to consider the possibility that growing or producing something to sell is actually not the only way to define sustainable farming. Granted, by definition, a farm enterprise can't be truly sustainable if it habitually loses money. But there's more than one way to look at "making" money. For example, suppose that instead of earning more money by selling more crops, we *save* money by growing or producing more of our own food?

My mother once said to me that there is a unique sense of empowerment in producing even a small proportion of your own food. It seems to me that the trend of "eating local," coinciding as it has with our national economic downturn, is a perfect opportunity — and should be a good motivation — for us all to do just that. It doesn't get much more local than growing your own. Also, considering that most of our food is transported quite some distance to us (see Chapter 36), it's clear that saving significantly on our food costs is a real possibility. And, of course, when we have even a small garden, we benefit from the freshness, great taste and nutrition of food grown seasonally.

What I'm suggesting is that we all take a thoughtful look at our lifestyles, the things that we value, our true priorities. Suppose we pass on to our children the joy and deep satisfaction of planting a seed, then nurturing and harvesting a crop? Suppose we redefine our financial lives by choosing to live debt-free? Suppose we start thinking in terms of the legacy of our choices, instead of just today's bottom line? Small farmers, backyard gardeners and city dwellers, take heart: Saving money and providing for yourself and your family are not just possible, they're great choices.

Think small, simplify. Buy local, sell local. Grow your own, share with others.

How's that for a legacy?

CHAPTER 37

Pure Poultry Premise #4: Have Fun!

I N CASE YOU WERE WONDERING, these *Pure Poultry* Premises are not being shared in any particular order. From our point of view, they are all of equal importance, since, as a group, they summarize our perspective on both farming and life (not that you can actually separate the two).

So, back to Premise #4. It's quite simple, actually: Have fun! Yes, that's right, *have fun*. The underlying goal of all that we are doing here is to create a life, and a farm system, that are sustainable, after all. And frankly, if it's not fun at least some of the time, why would you want to keep doing it? If your initial reaction to this is to ask why one should have to consciously plan to have fun, reread any random chapter of this book. It's not just that we're busy; who isn't these days? Our experience has been that it is too easy to get in a rut of just making it through today and hoping for some rest before getting up again at first light tomorrow. So we make a point of finding ways to bring some fun to the party.

Having fun isn't confined to off-work hours. The first year we had chickens, for example, we started a tradition of Happy Hour with the Chickens. During that summer, we had a couple of lawn chairs in the chicken yard, and we would take drinks (for us, you know) and some kind of treat for the birds. We had fun sitting down there among them for a while before we tucked them in for the night. It was lovely to relax for an hour or so and enjoy watching the Chicken Channel. Some of the bolder birds would jump up on our laps, depending on how yummy they judged that day's treats. They seem to be partial to things like organic ginger snaps and Tim's potato chips. Don't look at me like that; I don't make this stuff up.

It also makes a difference to me to vary details of the routine chores. Even reversing the order in which I make the rounds to pick up eggs seems to make it more fun. I have no idea why, but who cares? And it's not as if these chores ever feel particularly tedious anyway; between indoor and outdoor tasks, I get plenty of variety on any given day, provided I put some effort into planning my schedule.

Of course, there are plenty of ways to have fun outside the workday. It's good to have non-farm-related hobbies and interests, and getting involved in your community usually offers many choices for you social butterflies. We like to watch movies, David regularly goes out to social events with his friends, and we eat out at the Alder Wood Bistro, among other diversions.

I could go on, but I will leave it there, in the hope that you will use your imagination and start thinking up ways to regularly add fun to your own life. What could be more fun than going to bed each night actually looking forward to getting up the next day and doing it all again?

CHAPTER 38

Abundance and Gratitude

THERE ARE DAYS WHEN I FEEL ALMOST OVERWHELMED with a sense of the abundance in our lives. We're healthy, we have no debt, and we get to live in this amazingly beautiful place. Certainly, we face challenges like anyone else: in our marriage relationship, occasional unexpected financial issues and now and again being slowed down by illness or injury. I still struggle with ambivalence about raising animals to slaughter for food. I feel guilty when an ongoing physical challenge prevents me from doing my share of the work on the farm. It is not always easy, and not always fun.

But we keep going. Why? We love this place, we love our community, and doing what we do here has enabled us to contribute something to the quality of life of our customers, friends and family. Sure, having animals ties us down; we haven't been away together from the farm overnight since we got our first chickens in 2007.

The birds have contributed greatly to our lives, too. We've had a lot of fun over these years, observing and laughing over the different personalities, the funny and perplexing mannerisms and learning about the different calls they make and the predator-alert system. We have laughed ourselves silly over the goofy antics of the ducks and the shoe-fetish behavior of Old Tom. We have had many moments of frustration looking for nests of turkey hens who insist on laying eggs somewhere other than the perfectly nice broody coops I built for them. Many times I have cried when a bird was killed by a hawk or a bobcat, and felt guilty that I didn't do more to keep it safe.

Of course, there is more to the farm than chickens, turkeys and ducks. I love gardening, and it is positively delightful to have so much space and peat-rich soil available. Every year, though, I've cut down on the amount of planting I've done, always trying to find that elusive balance between all my responsibilities without becoming stressed or too tired. Our short growing season is a challenge. Over the past couple of seasons, I've managed to be organized enough to plant some fall and winter crops. Since my little greenhouse isn't heated, growing late-season crops seems to me to be a good way to effectively extend our gardening season well beyond the first October frosts.

Since our first year here, in the course of getting used to a shorter growing season, I had been wishing for a greenhouse to start plants earlier in the spring. A couple of years ago, and not a moment too soon, I had an inspiration. An old A-frame chicken coop (the first coop I built in 2007) had been sitting unused for some time, since our rapidly growing chicken flock had been moved into more spacious sleeping quarters. I stripped off the chicken wire and pulled out all the rusty staples, nails and hardware. Then I dismantled the frame just enough to make it easy for me to move up the hill to the spot I'd marked off for it at the north end of my kitchen garden.

My brother John had given us forty-two 8-by-8-by-16-inch concrete blocks. I made a sketch, did some number-crunching and discovered that I had just the right amount to build a two-course foundation for the 8-by-8-foot frame. I cleared and leveled the site and covered it with overlapping pieces of landscape cloth. Then I laid out the first course of blocks, taking my time and getting everything square and level. Once that was in place, the second course went up quickly. Lifting the frame onto the foundation was a little tricky. Excited about getting past the foundation stage, I was working on it one day when David was in Seattle. It would have been much easier with two people, but eventually I managed it. The frame was secured to the foundation with long masonry bolts.

David bought some clear polycarbonate panels to cover the greenhouse frame. They were just the right size so that, when overlapped with the next panel, four panels covered the sides nicely. I added a door on the south end and a window on the north; a nifty piece of hardware automatically opens the window when the greenhouse temperature gets above about 75°F. I installed two deep shelves on either side, thick cedar lumber wide enough to easily hold standard seedling flats.

Considering the footprint of my little greenhouse is only eight-feet square, it is surprising how much space there is for both storage and seed-starting. I also had the satisfaction of finding a new use for an unloved surplus coop, and the whole thing didn't cost much. Most of the cost was the polycarbonate panels (around $20 each, and I used ten), plus a little for hardware. It's made a huge difference to be able to start seeds much earlier in the year.

Every year is different; weather patterns, other projects that come up, family health issues — lots of factors affect what and how much we're able to grow. But I also work hard at preserving as much of the food we produce as possible, partly because I dislike wasting good

Coop converted to greenhouse.

food. I also have the additional satisfaction of a well-stocked pantry at any time of year.

Whenever we slaughter chickens, some inevitably end up in the stockpot. I make a nice big batch of chicken broth, strain it and process it in the pressure canner. Once the meat has been cooled and picked off the bones, it gets canned as well.

Sometimes, I'll combine broth with chicken meat, add a few garden vegetables and can it as chicken soup. But having the meat and broth in separate jars is handy too. Just the other night, when I knew David would be coming home late from visiting his mother in Seattle, I made chicken curry. I put a pan of his favorite organic brown basmati rice on to simmer, then got two pint jars of chicken and one quart jar of chicken broth from the pantry. I melted a stick of butter, sautéed some onions and garlic, then made a roux by stirring in a half cup of flour, a few grinds of black pepper, curry powder, a pinch of garam masala and a couple of pinches of cumin. Once the roux was lightly browned, I stirred in the broth. As soon as the mixture started to thicken, I dumped the two jars of chicken with its own broth into the pan. David likes to adjust the seasoning of dishes like curry with a little soy sauce instead of salt, and this time I thought of him and remembered the soy sauce. As brown basmati rice takes about forty-five minutes to cook, the curry was actually done well ahead of it. As I noted earlier, it's lovely not to have to thaw out anything. Plus, that batch of chicken curry provided three meals for us. You'll have to pardon me if I sound a little smug.

The Abundance of Having Less

When we moved from Seattle to the farm, we made a conscious decision to live debt-free. Even while living in Seattle, we had a mortgage but otherwise owed nothing. We drive used cars bought with cash and do much of the maintenance and repair work ourselves. We have no television, so we have no cable TV bill. Ditto for high-speed Internet service. Our culture has pushed the message for years that there is something wrong with us if we don't have the latest electronic gizmo, more cable channels, a newer car, a bigger house.

More, bigger. Bigger, more.

What about those of us who not only don't have all those things, but find that we are perfectly content living without them?

It seemed to me that, living in Seattle, even our relatively simple life was getting more complicated and more expensive all the time. Here, our lives are busy and the days full. (David likes to say that he went from working seventy hours a week to working one hundred, and he's not exaggerating.) But we have more choices available to us. We are not stuck in a cycle of working to maintain a certain lifestyle, just to maintain a certain lifestyle. We're growing more of our own food now. Would we like to be more self-sufficient? Sure. But we also see the need to be realistic. For one thing, we didn't start this new life when we were in our twenties.

Living off the grid has a way of changing your priorities. I truly thought I would miss having a television. I didn't spend hours every day with it, but I did enjoy watching a little news in the morning and *Law & Order* reruns in the evening. And although I've always been a big sports fan, about the only sports events I even think about anymore are the tennis grand slams. And honestly, I was surprised to find that I didn't miss the TV at all. These days, who has the time anyway?

I guess it comes down to knowing what is truly important to you and taking what steps you can to follow your heart. I've heard lots of people say, "Oh, I'd love to be off the grid." I suspect they might be as ignorant as I was about what that means. Possibly what they really are expressing is a desire to live a simpler life, away from city traffic and stress, closer to nature. We've also heard a few mildly disparaging remarks about us "going back to the land," and other comments that seem to imply that we're somehow strange. I am also taken aback when people react negatively to our desire to live debt-free. (We don't go out of our way to talk about this, but occasionally it does come up in conversation.)

Everyone, at some point in his life, has to decide which way he's going to go. We know our choice of lifestyle is right for us. It's not so convenient to have to take our garbage and recycling to the local dump instead of having curbside pickup like we did in Seattle. In the city, we could walk just about everywhere: to the grocery store, the local swimming pool, the public library; here we have to drive pretty much everywhere. There are always trade-offs, and I think that the majority in our situation are in our favor. There is an abundance of space, of trees and water and wildlife, of fresh air and starlight here that is precious to us. With all of the challenges this life has given us, we have so much to be grateful for and look forward to every day.

CHAPTER 39

On Being Unplugged

I TOTALLY GET THAT THIS IS NOT THE KIND OF LIFE that every woman dreams of. But I tell you, it works for me. I'm married to a man who shares my values and loves this place even more than I do. He brings with him the whole history of his family and the dreams that led them here nearly eighty years ago. I love the peace and quiet and the feeling of being surrounded by green, living, growing things. I also love the feeling of history here, of a time before these gorgeous trees were so large, before David's grandmother planted the old apple trees and the black walnut tree and the daffodils and lilac bushes that continue to burst into exuberant bloom every spring.

I often think about the time and skill and effort and love that David's grandfather put into building this solid, roomy, comfortable house. I wonder if Grandpa Moniz ever imagined one of his descendants living here one day, enjoying the same view of Blue Mountain when the first rays of morning sunlight warm its snowy face; the flock of red-winged blackbirds in the big cottonwood tree east of the bog, singing their little hearts out in the early-spring fog; the Great Blue Heron standing patiently, completely still, in the shallows of the pond, waiting for an unwary fish or frog to stray within reach of its enormous bill.

It might seem a stretch to say I think our birds love us, but I do believe they have had a happy life here. We have experienced so many

191

moments of laughter and wonder and joy in the course of caring for these beautiful birds. Sure, we've made plenty of mistakes, and it seems that we will never run out of new things to learn as we strive to deepen our understanding of the environment here, our role as stewards of the land and our animals and our responsibility to share what we have learned.

I love the way our heritage birds happily forage a good share of their food. They are excellent mothers and do a great job of hatching and raising their chicks so we don't have to buy new ones every year. They turn our compost, fertilize our gardens, gobble up weed seeds, slugs and bugs. They amuse us, amaze us, teach us and entertain us. They love the space, the pasture, the fresh air, spring water and organic grain. They reward us for our attentions with consistently beautiful, delicious, nutritious eggs, plentiful enough to share among our families and in our community.

These birds have contributed much more to our lives than just eggs, though. They've given us a new sense of purpose, not just in terms of our farm, but in the greater context of the community and the relationships that have grown out of our decision to raise heritage poultry.

In addition to not having cable TV and high-speed Internet, we have no cell phone reception here at the farm. When someone stays overnight for the first time, it's not unusual for them to be a little anxious. I think they probably are anticipating that it will be perhaps a bit on the primitive side. I don't blame them; before coming here, I really had no idea what being off the grid meant. And living here full-time is very different from coming up from the city for a night or two. The only real adjustment I remember having to make was getting out of the habit of reaching for a light switch whenever I came into a room.

When people plan to visit, we frequently have to assure them that they don't need to bring a sleeping bag, potable water or food. Just as frequently, our guests are pleasantly surprised at how comfortable they are here. I'm pretty sure they don't just mean the accommodations and food. Maybe I'm imagining it, but I swear I can see in their faces that the reality of being unplugged actually has nothing to do with deprivation. One old friend of mine, after spending the weekend here with his partner, called me to say that the two of them had reconnected while they were here. Apparently when at home, they typically spent lots of time watching television; since that wasn't an option here, they took walks in the woods together and long naps in the afternoons. We shared meals and after-dinner drinks by the fire. They helped me in the garden and loved picking up the eggs.

Contrary to what people often think, we did not deliberately seek out an off-grid home. It just so happens that the home that we live in is off the grid. We're so fortunate that we both have the temperament and the skills and abilities that fit so comfortably with being unplugged, not just on the occasional weekend, but every single day. I honestly don't miss anything about living in the city. I have family there, so I do visit. City life is just the thing for many people, and I think it's great that it works for them. And although I lived in Seattle all my life until moving to the farm, this is where I belong. I finally have that peaceful feeling of having put down my roots.

As I type this, I am looking out the north-facing kitchen window. A New Hampshire hen appears in the doorway of one of the coops, loudly and proudly announcing her accomplishment: another fresh egg huddles warmly in the nest box. A rooster and two hens are industriously scratching and pecking near the duck coops. Four Indian Runner ducks and a Khaki Campbell are happily dabbling in the muddy puddles left over from yesterday afternoon's rain. Another rooster is crowing, out of sight somewhere on the other side of the house. David has managed to sneak in time for a nap.

It's a good life, being unplugged. Our pantry is full; the wood stoves are burning merrily. My cat Cosmo is sacked out blissfully on his bed behind the living room wood stove. There is no through traffic anywhere near us, and everywhere we look there are trees. The pussy willows are out early this year. Our ponds are home to fish and frogs, lots of wild birds and the occasional river otter. We have a shooting range; I enjoy both shooting and archery. David and I both love to cook, and I make beer and wine and champagne. Yes, we're busy, but I find plenty of time to do the things I love, like relaxing in my chair by the wood stove with a good book every night.

All this, without any of that annoying 60-cycle electrical hum.

CHAPTER 40

Nesting

L ONG BEFORE WE MOVED TO THE FARM or even thought about raising animals, I had dreams of living in the country one day. While
living in Seattle, I regularly experienced the urge to travel. In the six
years between when we got married and David retired, we traveled
to Europe several times, as well as to Mexico, Montreal, California,
Nevada, Illinois and Indiana. Once we took a road trip east into
Montana, then north into Alberta and back home via eastern British
Columbia. At the end of these trips, I was always glad to be home, I
guess, but I really loved being on the road.

The funny thing is, ever since we moved to the farm, I haven't felt
the slightest urge to travel. We've been here seven years now, so I don't
think it's just a matter of being in a new place, going through the
transition of moving and adjusting to a completely different lifestyle.
It's more the sense of belonging, of not being here temporarily, of the
permanency of putting down roots and having room to grow.

There is a comfort, an ease in the daily routine with our birds.
When it gets light, we head out into the cool morning air, refill the
feeders and drinkers and, one by one, open up the coops. The ducks
tumble out in a rush and immediately race across the yard, stretching
and flapping their wings excitedly as if they haven't been outside for
days. The chickens are usually a little more sedate when they wake
up. They hop down from their coops one at a time and head over to

the nearest feeding station. The turkeys almost always move slowly, more like a saunter or an amble, after they descend from their roosts and casually stretch their wings and bodies skyward. The sun is up but hasn't come over the hills to the east yet. We pick up the duck eggs and head back inside for a while.

A couple of times during the day, I usually check the birds' food and water supply and pick up the chicken eggs. Eggs are washed, dried, weighed, packed in cartons according to size and refrigerated. Clean dry bedding is added to coops and nest boxes, and once a week, various coops are cleaned out as needed.

And then there's the rest of the routine: Clean house, file receipts, do laundry, cook, wash dishes, make lists, buy groceries, pay bills, file, sort, organize, pick up, dust, sweep, scrub, pump water. Feed the birds, pick up eggs, send out egg invoices, order more feed.

Naturally the routine varies somewhat depending on the season. But overall there is the knowledge that certain things will be the same tomorrow, as solid as the mountains always looking over our shoulders. And at the end of the day, we know that we did our best, once again, to see that our birds have had another good day. We're so happy and grateful that they are a part of this life that we have chosen and love.

CHAPTER 41

Tomorrow

MAY 26, 2007. It's hard to believe it's been less than six years since Chicken Day, when we got our first chickens. Looking back, it's amazing how quickly things changed for us at that point. In less than a year, we had rapidly built up the size of our flock and had lots of extra eggs. This led to getting our egg dealer's license and selling eggs to the Alder Wood Bistro. That same spring, we got our first turkeys. And ducks. We're still wondering just what makes our eggs so delicious. Is it just a matter of being so fresh and organic? Could it be our calcium-rich, unchlorinated mountain spring water? Or is it their penchant for foraging on pasture all day, grazing on green leafy things and snatching up all manner of creeping and crawling things?

Most likely it is a combination of factors. We often hear people suggest that it's because our eggs "come from happy hens." For quite a while, I simply dismissed that theory with a somewhat condescending smile. The more I thought about it, though, the more I wondered if that could be the answer. Certainly we try hard to raise our birds in a way that allows them as much of their natural behavior as possible, while keeping them safe. They have plenty of room to move around and fresh air all day. They make nests, incubate eggs and hatch and raise their offspring. They roost, fly, dust-bathe and mate. They bicker among themselves about the pecking order.

While plenty of potential predators are around, the birds also have access to lots of protective cover. There are trees, bushes, even cars for them to run under when the aerial-predator alarm is sounded.

What's not to be happy about? But how would being happy affect the quality of the eggs? I don't know. My best answer at the moment is that it is simply a mystery and miracle of nature.

I have talked a few times about the importance of planning. To my mind, it is equally important to periodically evaluate and see how things are going. We have had several discussions recently about the possibility of making some changes. For example, should we continue to raise turkeys? There is no simple answer. It depends on what our goals are now, what our local community wants, what our farming friends are up to and a few other factors. As with many issues, I have decidedly mixed feelings about keeping turkeys. I don't think this is a bad thing, necessarily, as long as my ambivalence motivates me to take an honest look at the situation and choose the option that makes the most sense in relation to the big picture. Heritage turkeys are sweet. They're often funny. Although the quality of their meat is outstanding, we figured out long ago that, unless we were raising nothing but turkeys, it's almost impossible to make raising them for slaughter cost-effective in our location. We live hours away from the nearest USDA-certified facility that processes poultry, and we can't sell them wholesale if we slaughter them on the farm. The most we can do is sell directly to retail customers, and the customers must come to the farm to buy them.

We also wouldn't want to raise only one kind of livestock, poultry or anything else. We firmly believe that diversity is key to the success of any small farm. I mentioned in an earlier chapter that ducks follow pigs after they have plowed up their paddock. Ducks also clean up slugs, which, for the most part, chickens and turkeys don't. A point in favor of turkeys is their valuable predator-alert skills. The roosters are pretty good at spotting aerial predators and sounding the alarm, but turkeys are hands-down superior for letting everyone, including us, know when a potential threat is on the ground. And turkeys, once they are approaching adulthood, are large enough not to have many predator concerns of their own.

So we're grappling with the turkey question. It may sound as if our main issue with them is that we're not making money. You should know by now that keeping poultry is not simply a matter of money for us. It is, however, a matter of economics. Remember the unromantic truth about farming: If a farm enterprise is not earning money or even paying for itself, it is by definition not sustainable. And while we might be able to think of other reasons we ought to keep at least some turkeys around, I think it's irresponsible to completely ignore the financial side of everything we're doing here.

As for the ducks and chickens, just what is the optimum size of our laying flocks? Obviously that depends on how many customers we have and how many eggs they need. Since our main customer is a restaurant, however, it becomes a tricky question to answer because of its seasonal fluctuations. In 2012, we were able to sell our extra eggs at local retail stores during the spring. In the summer, its busiest season, the Bistro bought most of our eggs, so we had fewer eggs to sell elsewhere. In the fall, when the restaurant's business is slowing down, so is the egg production; the birds moult and all but stop laying for a couple of months. My guess is that the "perfect" flock size will always be a bit of a moving target. You never know when you'll lose a few birds in a sudden rash of hawk attacks, such as we experienced in 2012. There's no guarantee how many chicks will be hatched

and survive, or how many of those will be laying hens or ducks. My feeling is that keeping production and sales records (which we do anyway because it's required for our organic certification) and periodically reviewing the numbers will help us to spot trends earlier and make adjustments where and when we can.

And of course, we're both getting older. Right now we're both in pretty good shape, able to cut down trees, haul, split and stack firewood to heat our home. But we are aware that, physical issues aside, we may very well come to a point of simply not wanting to work this hard to heat the house anymore.

So we have a plan B in place, at least for heating the house. In 2007, when all the gas lines in our house were replaced, one was installed right outside our living room, straight across from our wood stove. At any time, we can replace the wood stove with a gas stove or heater, punch a hole in the wall and connect it to the gas line. We already have a gas stove and oven in the kitchen, in addition to the wood-burning cook stove, so cooking isn't an issue. Whether one gas stove or heater in the living room will be sufficient to heat the entire house remains to be seen.

David and I haven't always agreed on everything about our poultry enterprise. Actually, he tends to describe what we do as "a hobby gone bad." And I know what he means; it's more of a labor of love in many ways. I know it's really a business, but I look at it differently because I keep the records and do the taxes. We've always agreed, though, that, as long as we have birds or other animals up here, we're going to continue to do our best to give them the best life possible. And when necessary, they will be slaughtered with all the dignity they deserve and the respect that we continue to feel for them.

So what's next? There is always a list of projects. We can't get to everything every day, so we have to prioritize. I've sometimes thought if only we had no birds here for a year or two then we could get some other things done. Probably the reality is that, even if we had no birds, we'd find ourselves having to pick and choose which projects to start and which to shelve. How we prioritize depends on our cash flow, our busy seasonal work, the shorter days and uncooperative weather.

And what about having the time to travel? Even though traveling is not so important to me now, it would be nice to be able to simply pack an overnight bag and take off for the weekend sometimes. I'm still trying to learn to be flexible, although I've gotten somewhat better at it. It also requires patience, willingness to negotiate and a readiness to not take any of it too terribly seriously.

Will we continue to raise pigs part of each year? I don't know. As with so many questions, there is no obvious, clear-cut answer; there are always pros and cons to consider.

As I write this, Cosmo is curled up fast asleep (again) on a sheepskin rug near the wood stove. David is down the hill in town, delivering eggs. Our chickens, turkeys and ducks are roaming around outside, scratching and pecking through the melting January snow. Breeding season and gardening season are right around the corner. Like Christmas, they always come the same time of year, and always seem to sneak up on me anyway. In a couple of hours, the birds will be heading into their coops for the night, and we will walk around, closing and latching the coop doors.

So much of what we do every day is pure routine, expected, normal and familiar. Yet there are also situations that come up: an unfamiliar predator, difficult weather conditions, an inexplicable drop in egg production, a day when one of us feels overwhelmed by the chores that must be done before dark. From the moment when I open the coop doors in the cool of the morning, watching these beautiful birds begin another day, my hope is that I will learn something today, and will continue to learn. I know for sure that, later this evening, when I put my book down, turn out the gas lamps and head up to bed, I will be looking forward to doing it all again tomorrow.

Appendix A:
Poultry from Scratch Worksheet

IN EACH SECTION, circle all answers that apply. Add additional information where needed.

1. **General**

   I am interested in raising:
   - Chickens
   - Turkeys
   - Ducks
   - Other _____

   I want to raise poultry because:
   - Pet or 4-H project
   - Meat and/or eggs for my family
   - To make money
   - Novelty, just to try something different
   - To breed and/or show
   - Other _____

   My family and home situation is:
   - I have a job outside of my home (PT or FT?) and am away _____ hours per day.
   - I have _____ children at home, of whom _____ are old enough to help with birds.

• I have dogs or other outdoor pets and/or livestock (list type and number of each).

_____

_____

_____

_____

Other family-related concerns or questions that occur to me:

_____

_____

_____

2. Time

• I have plenty of time to spend learning about and caring for my poultry.
• I have limited time due to (circle) my job, family issues, other animals.
• Other time-related questions and concerns: _____

_____

_____

3. Space

• I have plenty of room for my birds to roam or free-range during the day.
• I have limited space and will need to confine my birds to a coop or pen.
• I plan to free-range my birds, and understand how many birds I can keep on the amount of pasture or yard space I have.

_____

_____

_____

_____

• Other space-related questions and concerns: _____

_____

_____

4. **Housing**
   - I have already prepared (circle all that apply) housing, roosts, nest boxes, fencing.
   - I have a good supply of appropriate bedding for the coop(s).
   - I understand how often I need to clean out the coop(s) and have a plan for this.
   - My coop(s) have adequate ventilation.
   - Other housing-related questions and concerns: _____

   _____

   _____

5. **Feed**
   - I understand what feed is appropriate for each stage of my birds' growth stages.
   - I have located a nearby source of the correct feed (Name, hours and contact info): _____

   _____

   _____

   _____

   - I haven't yet located a reliable source of the feed I want.
   - Other feed-related questions and concerns: _____

   _____

   _____

6. **Health**
   - I am willing to spend some time learning about common poultry health issues and take appropriate preventative measures (list): _____

   _____

   - I know of a local veterinarian who handles poultry (Name, hours, contact info): _____

   _____

   _____

   _____

- I have a plan for dealing with sick or injured birds who may be suffering: _____
  _____
  _____

- Other health-related questions and concerns: _____
  _____
  _____

7. Breeding

- I want to breed poultry mainly to replenish my flock each year.
- I want to breed poultry seriously to help preserve a rare breed or to show.
- I have prepared additional housing and nest box space needed for incubating hens.
- I plan to collect hatching eggs and incubate them artificially.
- I am prepared to brood chicks if a hen should abandon her nest.
- I have a plan for dealing with extra roosters, inferior birds and other cull issues: _____
  _____
  _____

- Other breeding-related questions and concerns: _____
  _____
  _____

8. Predators

- Possible predators in my area include:
  Owls
  Hawks and eagle
  Coyotes
  Dogs
  Cougars
  Bobcats
  Weasels or mink

Skunks

Raccoons

Rats

Other _____

- As far as I know, there are no potential predators near my poultry.
- My plan for minimizing or preventing daytime attacks is: __

- My plan for minimizing or preventing nighttime attacks is:

  _____

  _____

- Other predator-related questions and concerns: _____

  _____

  _____

9. **Egg birds**
   - There are _____ egg eaters in my family. I plan to keep _____ laying hens to supply enough eggs for us.
   - I'm thinking of raising a larger laying flock and selling the extra eggs.
   - I am aware of local and state licensing laws related to selling eggs. (Describe) _____

     _____

     _____

   - I don't want to sell eggs, but I have a plan for dealing with extra eggs when there is a surplus. _____

     _____

     _____

   - I have a plan for dealing with older hens who are past their prime laying years. _____

     _____

     _____

   - Other egg-related questions and concerns: _____

     _____

     _____

10. Meat birds

- I plan to raise _____ meat birds per year for my family.
- My preferred breed for meat birds is: _____

  _____ .

- I know of a nearby source of appropriate meat bird feed
  (Name, hours and contact info): _____

  _____

  _____

- I own or have access to appropriate slaughtering equipment.
- I know someone with slaughtering experience who is willing
  to show me the ropes.
- I have decided which method of slaughtering meat birds is
  preferable for me: _____

  _____

  _____

- I plan to raise fast-growing hybrid meat birds, and am confi-
  dent I can get them slaughtered in a timely manner.
- I have plenty of freezer space to store _____ whole birds.
- I want to raise meat birds to sell, and am aware of local and
  state licensing regulations and facility requirements.
- Other meat-related questions and concerns: _____

  _____

  _____

Appendix B:
Resources

Books

IOWN AND REGULARLY REFER TO EVERY SINGLE ONE OF THESE BOOKS. I have deliberately listed them in no particular order; it's not a very long list, and I thought I'd try to encourage you to look it over and pick out a few.

Turkey Production, L.E. Cline, Orange Judd Publishing, 1st Edition, 1929.
It might be a challenge finding a copy of this book, but in my opinion it is well worth the effort, and well worth reading.

Chicken Tractor, Andy Lee and Patricia Foreman, Good Earth Publications, 3rd Edition, 2011.
A positively delightful, as well as insightful, read.

Day-Range Poultry, Andy Lee and Patricia Foreman, Good Earth Publications, 2005.
More useful and entertaining information by the authors of *Chicken Tractor*.

Storey's Guide to Raising Ducks, Dave Holderread, Storey Publishing, 2001.
Excellent all-around handbook by an acknowledged expert on raising waterfowl. Highly recommended.

All Flesh Is Grass: The Pleasures and Promises of Pasture Farming, Gene Logsdon, Swallow Press/Ohio University Press, 2004.
Gene Logsdon is one of my very favorite writers. This is a wonderful guide to raising poultry and other livestock on pasture. It includes descriptions of many pasture plants.

Charcuterie: The Craft of Salting, Smoking and Curing, Michael Ruhlman, Brian Polcyn and Thomas Keller, Norton, 2005.
Lots of good recipes and techniques for curing meat and making sausages. Some conflicting information on the subject of using nitrates, but otherwise a very good book that I use regularly.

How to Build Animal Housing, Carol Ekarius, Storey Publishing, 2004.
Fabulous handbook for building all kinds of housing and shelters for livestock, from the smallest nest box or rabbit hutch to a great big beautiful milking barn. Includes essential information about minimum space requirements for different animals, and much more.

Storey's Guide to Raising Chickens, Gail Damerow, Storey Publishing, 3rd Edition, 2010.
Classic and very useful reference for raising chickens, from hatching through slaughtering.

The Farmstead Egg Cookbook, Terry Golson, St. Martin's Press, 2006.
I love this little book. Lots of recipes and ideas for ways to use and showcase those farm-fresh eggs.

Ball Complete Book of Home Preserving, Judi Kingry and Lauren Devine, Eds., Robert Rose, 2006.
Handy reference book with many interesting recipes for canning and preserving. It doesn't have much on pressure canning, but what canning book does?

Storey's Illustrated Guide to Poultry Breeds, Carol Ekarius, Storey Publishing, 2007.
You might buy this book for the gorgeous photography alone, but you'll want to read the articles too. Extremely helpful if you're trying to decide which poultry breeds might work for you.

Homegrown and Handmade: A Practical Guide to More Self-reliant Living, Deborah Niemann, New Society Publishers, 2011.
An amazing amount of information all in one book by a popular expert in self-sufficient living. Easy to read and understand, and truly inspiring.

The Chicken Health Handbook, Gail Damerow, Storey Publishing, 1994.
An essential reference to have on your shelf. Diagnostic charts, excellent information on common parasites, chicken anatomy and much more.

Humane Livestock Handling, Temple Grandin and Mark Deesing, Storey Publishing, 2008.
Fascinating and inspiring book. I recommend it for anyone raising any kind of livestock, especially if you ever have to move or transport animals.

The Nankin Bantam: A Rare and Ancient Fowl, Mark A. Fields, American Livestock Breeds Conservancy, 2006.
Good reference on this critically endangered bantam chicken, including its interesting history.

The Mating and Breeding of Poultry, Harry M. Lamon and Rob R. Slocum, Lyons Press, 2003, originally published in 1920.
The printing is small and often hard to read, but this is still a useful book to have on hand if you're considering a serious poultry-breeding program.

Making Mobile Hen Houses, Michael Roberts, Gold Cockerel Books, 2004.
A little book with lots of creative plans for building mobile chicken coops of different sizes. Published in England, so you'll have to make adjustments with the lumber sizes, but easy to follow.

Ducks and Geese in Your Backyard: A Beginner's Guide, Rick and Gail Luttmann, 1978.
The first book I had on raising ducks. I still have it and use it, along with *Storey's Guide to Raising Ducks*. Well worth it if you can find a copy.

Animals Make Us Human: Creating the Best Life for Animals, Temple Grandin and Catherine Johnson, Houghton Mifflin Harcourt, 2010. Another wonderful book by the author of *Animals in Translation*. Covers livestock, cats, dogs and more; the chapters on wildlife and zoos are quite fascinating. If you think you know a lot about how animals should be treated, get ready to learn some more.

The Omnivore's Dilemma: A Natural History of Four Meals, Michael Pollan, Penguin, 2006.
More food for thought from Michael Pollan on America's food industry. He diligently traces the ingredients of a typical fast-food meal back to their origins, uncovers some surprising facts about the organic food business and puts together his own hunter-gatherer meal, foraging and hunting nearly all the ingredients himself. An entertaining — and sobering — read.

Internet Resources

As you can see, this is a very short section. I don't spend much time on the Internet myself, partly because we don't have full-time or high-speed service, mostly because I just don't have the time. Here are a few Websites that I've found to be useful on the subject of heritage poultry.

The Pot Pies and Egg Money blog
http://potpiesandeggmoney.blogspot.com
OK, so I don't find this blog useful so much as a forum in which to air my own experiences, questions and opinions about just about anything that happens on our farm. Photos of our farm, birds and pigs, too!

Backyard Chickens
www.backyardchickens.com
A nicely designed, extensive Website with forums on just about any poultry-related subject you can think of.

Farm Forward
www.farmforward.com
I just recently came across this site. The home page states: "Farm

Forward implements innovative strategies to promote conscientious food choices, reduce farm animal suffering, and advance sustainable agriculture." Good enough for me.

The Contrary Farmer blog

www.thecontraryfarmer.wordpress.com
Gene Logsdon's wonderful and witty take on everything about farm life. I don't subscribe to that many blogs, but this one is definitely worth reading.

Businesses and Organizations

Alder Wood Bistro

139 W. Alder St., Sequim, WA 98382
(360) 683-4321
www.alderwoodbistro.com
If you're in the Sequim area, you really must make time to have dinner at the Bistro. Locally sourced, seasonal, organic food; fresh Northwest cuisine at its absolute best. Daily specials and wonderful pizzas from the custom-built wood-fired oven. Open for dinner Tuesday through Saturday, 4:30-9:00 PM. Reservations recommended.

In Season Farms

(604) 857-5781
E-mail: organicfeed@shawbiz.ca (no Website as of this writing)
Our long-time supplier of organic feed for our birds and pigs. In Season has an extensive list of available organic feed formulas, as well as cracked corn, split peas, kelp meal and other supplements. Although they are located across the Washington border in British Columbia, In Season delivers feed over a wide area; they come our way every three weeks. Buy the 20-kilo bag or get the half- or full-ton tote. If you're in western Washington, you can also find their feed by the bag in some retail feed stores. Price list available upon request.

Humane Society of the United States

http://www.humanesociety.org/issues/confinement_farm/facts/guide_egg_labels.html

Up-to-date information about terms used to label poultry eggs and how these relate to the welfare of production hens.

Manini's
www.maninis.com
A Seattle-based company that makes wonderful gluten-free bread, pasta and baking mixes. I like these because they are formulated with "ancient grains," not rice. I regularly make their gluten-free fresh pasta, and it is delicious. All their mixes are available in 5-pound or 50-pound bags.

Woodstock Soapstone Company
www.woodstove.com
Woodstock makes beautiful EPA-certified soapstone wood stoves and gas stoves. We bought one of their wood stoves right before moving to the farm. It is gorgeous, efficient and features a catalytic combustor that minimizes harmful emissions. We love this company!

Territorial Seed Company
www.territorialseed.com
Not the cheapest place to buy seeds, but they have more organic seed choices every year, and a selection of grains and pasture plant seeds as well.

Bountiful Gardens
www.bountifulgardens.org
One of my very favorite seed catalogs, from the organization founded by John Jeavons. Good selection of grain seeds, although many are only available in small quantities. One of those catalogs you can learn a lot from even if you don't buy anything.

American Livestock Breeds Conservancy
www.albc-usa.org
You really should think about joining this non-profit organization. Connect with breeders of heritage poultry and other livestock and find all kinds of useful husbandry information, from the people who really know and advocate the preservation of our heritage breeds.

ATTRA: Appropriate Technology Transfer for Rural Areas

(800) 346-9140

www.attra.org

A federally funded program that offers publications and technical assistance in both production and marketing.

Chef's Collaborative

(617) 236-5200

www.chefscollaborative.org

A national network of over 1,000 culinary professionals who promote sustainable cuisine by celebrating the joys of local, seasonal and artisanal cooking. The organization's mission is to provide education and helpful tools that encourage local and sustainable food purchasing.

Farms Oceans Ranches Kitchens Stewards (FORKS)/ Chef's Collaborative Affiliate

E-mail: forkscontact@hotmail.com

www.forksproject.org

The Seattle Chapter of the Chef's Collaborative, focused on educating Northwest food system stakeholders about sustainable food system practices.

The Food Alliance

(503) 493-1066

www.foodalliance.org

This non-profit organization promotes sustainable agriculture by recognizing and rewarding farmers who produce food in environmentally friendly and socially responsible ways. It also strives to educate consumers and others in the food system about the benefits of sustainable agriculture.

North American Farmers' Direct Marketing Association

(888) 884-9270

www.nafdma.com

The NAFDMA is a great place for family farmers, extension agents and farm market managers to network with each other on the profitability of direct marketing. Members increase their farm income by learning from each other through conferences, international farm tours, newsletters, workshops and trade publications.

Slow Food USA
(212) 965-5640

www.slowfoodusa.org

This international organization is dedicated to the preservation of traditional food production and preparation, enhanced biodiversity and the revival of the kitchen and table as centers of pleasure, culture and economy.

WSDA Small Farm & Direct Marketing Program
(360) 902-1884

E-mail: smallfarms@agr.wa.gov

www.agr.wa.gov/Marketing/SmallFarm

This program assists farmers to understand current marketing regulations, addresses barriers in marketing regulations, assists in developing infrastructure necessary to market farm products and farmers markets, promotes localized food systems and provides a voice for small-scale agriculture within State government. In Washington State, this is the organization to contact for an egg dealer's license, which is required for selling eggs wholesale.

Washington State Farmers Market Association
(206) 706-5198

www.wafarmersmarkets.com

This non-profit network of over 80 farmers markets across the state is dedicated to working with other agricultural groups and agencies to provide workshops and marketing resources and to produce the annual Washington State Farmers Market Directory.

Index

About the Author

Victoria Redhed Miller grew up in Seattle, in a family of eight children. She married her husband David in 2000, and in 2006, they moved to their off-grid homestead in the foothills of Washington State's Olympic Mountains. The property had been bought by David's grandparents in 1936, and moving there to live fulfilled one of his childhood dreams.

In 2007, they began raising heritage-breed chickens, joined in 2008 by turkeys and ducks. Within a year, they obtained an egg dealer's license and have been selling chicken and duck eggs to a local restaurant and several retail stores ever since.

Both skilled do-it-yourselfers, Victoria and David designed and installed their solar electric system, and have plans for hydroelectric power as well. Victoria loves photography, reading, gardening, canning, meat curing and making beer and wine. Most recently, she has built a licensed microdistillery on their property.

If you have enjoyed *Pure Poultry* you might also enjoy other

Books to Build a New Society

Our books provide positive solutions for people who want to make a difference. We specialize in:

**Sustainable Living • Green Building • Peak Oil
Renewable Energy • Environment & Economy
Natural Building & Appropriate Technology
Progressive Leadership • Resistance and Community
Educational & Parenting Resources**

New Society Publishers

ENVIRONMENTAL BENEFITS STATEMENT

New Society Publishers has chosen to produce this book on recycled paper made with **100% post consumer waste,** processed chlorine free, and old growth free.

For every 5,000 books printed, New Society saves the following resources:[1]

24	Trees
2,192	Pounds of Solid Waste
2,412	Gallons of Water
3,146	Kilowatt Hours of Electricity
3,985	Pounds of Greenhouse Gases
17	Pounds of HAPs, VOCs, and AOX Combined
6	Cubic Yards of Landfill Space

[1]Environmental benefits are calculated based on research done by the Environmental Defense Fund and other members of the Paper Task Force who study the environmental impacts of the paper industry.

For a full list of NSP's titles, please call 1-800-567-6772 *or check out our website* at:

www.newsociety.com